# Tales of Discovery

Miguel M. Garcia
Editor

# Tales of Discovery

Delving into the World of Biology and Medicine

Springer

*Editor*
Miguel M. Garcia (ID)
Area of Pharmacology
Nutrition and Bromatology
Department of Basic Health Sciences
Unidad Asociada de I+D+i al Instituto de
Química Médica (IQM-CSIC)
Universidad Rey Juan Carlos
Madrid, Spain

*Illustrations by*
Raquel Sardá Sánchez
Area of Sculpture, Department of Arts
and Humanities
Universidad Rey Juan Carlos
Aranjuez, Madrid, Spain

María Martínez de Ubago Campos
Area of Drawing, Department of Arts
and Humanities
Universidad Rey Juan Carlos
Aranjuez, Madrid, Spain

ISBN 978-3-031-47619-8      ISBN 978-3-031-47620-4   (eBook)
https://doi.org/10.1007/978-3-031-47620-4

The translation was done with the help of artificial intelligence (machine translation by the service DeepL.com). A subsequent human revision was done primarily in terms of content.

This Springer imprint is published by the registered company Springer Nature Switzerland AG
The registered company address is: Gewerbestrasse 11, 6330 Cham, Switzerland

Paper in this product is recyclable.

# Foreword

The Oxford English dictionary gives six meanings for the definition of *story*:

1. A description of events and people that the writer or speaker has invented in order to entertain people;
2. The series of events in a book, film, play, etc.;
3. An account of past events or of how something has developed;
4. An account, often spoken, of what happened to somebody or of how something happened;
5. A report in a newspaper, magazine, or news broadcast;
6. Something that somebody says which is not true.

This same dictionary gives four meanings for the definition of *science*:

1. Knowledge about the structure and behavior of the natural and physical world, based on facts that you can prove, for example by experiments;
2. The study of science;
3. A particular branch of science;
4. A system for organizing the knowledge about a particular subject, especially one that deals with aspects of human behavior or society.

In the work presented here, these two terms (*stories* and *science*), apparently distant but nevertheless very close, merge.

Science is exciting, so is reading a story. Discovering and explaining the origin of things, phenomena, or events that occur around us is not a simple task. There are many observations that have to be made. The complicated interrelationships between the observations have to be explained. Sometimes, methodologies have to be established to be able to verify that everything observed and deduced is true. One of the primary characteristics of the long road in the scientific method is this: it is exciting! The same thing happens with a good story—a compelling tale. We enjoy it—and sometimes we can get lost in it.

These short stories of science are structured into sixteen short scientific tales which show us, in a simple way, theories about the origin of life, our relationship with bacteria, the ecological footprint of some ruminants, the secret life of fungi, the consequences of current lifestyle trends, the theory of aging, food addiction, or the treatment of pain, the fifth vital sign.

The work is written by 20 researchers from different academic fields (biology, pharmacy, medicine, and food science and technology), whose professional work includes teaching, research, and scientific dissemination. These varied aspects add value to the work and make it easy to understand and enjoyable reading, while maintaining the rigor of the information presented. The illustrations, created by professors of Fine Arts, deserve special mention. They are completely original works and reflect their scientific theories with total accuracy. Science and Art. Art and Science. Another exciting fusion!

This book reveals the exciting world of science through entertaining short stories, without losing scientific rigor. They demonstrate the contribution research makes to our understanding of the world around us, but also the significance of research to improving such world.

Visitación López-Miranda González ⓘ
Area of Pharmacology
Nutrition and Bromatology
Department of Basic Health Sciences
Unidad Asociada de I+D+i al Instituto
de Química Médica (IQM-CSIC)
Universidad Rey Juan Carlos
Madrid, Spain
visitacion.lopezmiranda@urjc.es

High Performance Experimental
Pharmacology Research Group
(PHARMAKOM)
Universidad Rey Juan Carlos
Alcorcón, Spain

Grupo Multidisciplinar de
Investigación y Tratamiento del Dolor
(i+DOL)
Alcorcón, Spain

# Preface

This book is intended primarily for use by anyone with science interest but also by students with a background in science. In addition to the text, all figures are original and have been conceived after thoroughly studying and discussing each topic with the corresponding authors. Each chapter (emplaced in the fields of genomic, endocrinology and metabolism, pain, food and environmental industry, cell aging, addiction and therapeutics or evolutionary biology) begins with a short introduction, followed by separate subtitles that introduce projecting issues from the main topic.

As science is becoming more and more specialized, academic and research institutions are demanding professional growth on teaching, research, and transference. At the same time, the European common framework advocates for multidisciplinary and multicentric works, which deepen in that specialized profile, but adopting a holistic focus. Under this frame, divulgation of research is although accessible, yet too specialized and incomprehensible for many. Besides, evenly mastering all three skills is an enormous task. For some people, science has made an impression on their consciousness, but it has yet not provided a good reason to the taxpayer to accept an increase of the budget set aside for research aims.

In addition to this, through the past years we have noticed that many of our students lack the practical application for the knowledge base they amass. Our research is often far from what we teach, and at the same time, our laboratories are made up of people specialized on a particular purpose or area of knowledge but with different backgrounds: physicians, dentists, physical therapists, biologists, biochemists, pharmaceutics or food and science technologists, just to mention a few. All in all, this favors the presence of education inequities and gaps in a milieu that is eminently biological. No one is devoid of areas of ignorance and confusion. The university entourage and research career provide however a magnificent and diversified contact portfolio. After having taken part in multiple science festivals and events to discuss our latest research with local average citizens, we decided to capture our talks in a book series to reach more people. It is with this aim that *Tales of Discovery: Delving into the World of Biology and Medicine* has been created.

The book aims to present current topics on varied disciplines approaching them to the ordinary reader in a language that they can understand, avoiding the technicality of an academic journal but also the personal tone of a speech or blog and produced in a media that insures rigor and reliability. Made up mainly of postdoctoral researchers and assistant professors with current or past positions in different locations worldwide, the group of scientists herein aims to let the reader take a glance at some of the research we perform and the concepts and the questions we deal with in our daily routine. Furthermore, I strongly believe that effective communication is key to succeed. Pursuing this goal, we have turned the tables and illustrations, normally performed by biologists with artistic flair, have been carried out by two colleagues from Fine Arts, experts in digital illustration, drawing, and design. The illustrations herein have the advantage that they have been designed by strangers to the biological field, in equal conditions with the ordinary citizen. In this sense, particular stress has been laid upon the artistic and conceptual meaning of the illustrations. It is my wish that the book presents a balance between scientific correctness and reading comprehension for the largest audience possible.

During the writing and compilation of this book, I have received the help and advice of my masters, professors Carlos Goicoechea and Visitación López-Miranda, who transmitted me their conviction on the richness of this enterprise. Together with the authors themselves, they were first to believe in this project. Life is way easier with comforting words and encouragement messages. Special recognition deserve professors Raquel Sardá and María Martínez de Ubago, responsible for the art concept, idea, drawing, and digital illustration, whom I started this project hand in hand with. They have been repeatedly meeting every author to study the fundamentals of the chapters and to understand what to capture on the different illustrations, spending endless hours also at night and during their holidays. I would also like to express my most sincere thanks to Gonzalo Córdova, Scientific, Publishing, and Commissioning Editor at Springer Nature. Sometimes you just need someone with a shared view who can understand and envisage the same outlook as you do. Gonzalo has been that bit of an oracle to me. He took a stance for us from the first moment, notwithstanding the role of Springer Nature as a world leader publisher on specialized titles. He had the patience and presence of mind to await the finalization of the project, no matter what. Last but not least, I would like to extend my special thanks to Chris Clarke, who not only did translate the chapters from a preliminary broken English, but interpreted our words gracefully and conferred a homogeneous narration to the work. It must certainly have been an arduous work striving with the language in addition to the intricacy of the topics.

Madrid, Spain                                                                                    Miguel M. Garcia

# Contents

# Editor, Illustrators and Contributors

## About the Editor

**Miguel M. Garcia** holds a degree in Biochemistry from Universidad Autónoma de Madrid (UAM), a master's degree in the Study and Treatment of Pain, and a Ph.D. in Pain Research from Universidad Rey Juan Carlos (URJC). He is currently Assistant Professor and Researcher in the Area of Pharmacology, Nutrition and Bromatology of the Department of Basic Health Sciences at Universidad Rey Juan Carlos. He belongs to the High Performance Research Group in Experimental Pharmacology (PHARMAKOM) and coordinates the Teaching Innovation Group in Diseases and their Treatment (EducaPath) of Universidad Rey Juan Carlos. He is a member of the International Association for the Study of Pain (IASP), the Spanish Pain Society (SED), and the Spanish Society of Pharmacology (SEF). His research work focuses on basic pharmacology in the field of pain, particularly on the role of glial cells and cannabinoid and TLR4 receptors in nociception. As a disseminator, he has experience as a speaker at different science festivals and activities.

## About the Illustrators

**Raquel Sardá Sánchez** holds a degree in Fine Arts from Universidad Complutense de Madrid (UCM) and a Ph.D. in Communication Sciences from Universidad Rey Juan Carlos (URJC), where she has been working since 2004, and is currently Associate Professor. She coordinates the degree in Fine Arts, the Teaching Innovation Group in Art and Technology: Methodologies, Strategies and Means (ArTec), and is Director of the University Master's Degree in Contemporary Artistic Practices. She also makes part of the Consolidated Research Group in Visual Culture and Contemporary Artistic Practices: Languages, Technologies and Means (CUVPAC) of Universidad Rey Juan Carlos. Between 1997 and 2008, she worked for the company *Telefónica* as Designer and Director of Multimedia Projects and Technological Innovation applied to art and heritage conservation. She has participated in more than forty national and international teaching innovation conferences and has taught at universities in Turkey, Germany, Italy, Belgium, France, Finland, Canada and China. Her artistic projects have been exhibited at the *Robert Capa* *was here festival* (MNCARS, Madrid), Cibart Festival 2017 (Seravezza, Italy), *The First Dream of Sunbird* (Yiwu, China), *Hans Kock Symposio* (Seekamp, Germany), *Fundación Montenmedio de Arte Contemporáneo* (Cádiz, Spain), the International Exhibition of Contemporary Medal (Ottawa, Canada; Sofia, Bulgaria), and at the *Real Jardín Botánico de Madrid*.

**María Martínez de Ubago Campos** holds a degree in Fine Arts, specializing in Sculpture, and obtained her master's degree in Art, Creation and Research and her Ph.D. in Fine Arts from Universidad Complutense de Madrid (UCM). She currently works as Researcher and Assistant Professor in Fine Arts and Fashion Design and Management at Universidad Rey Juan Carlos (URJC) and makes part of the Teaching Innovation Group in Design, Innovation, Art and Fashion (DINAMO). Her lines of research are focused on educational innovation with the use of art as a learning tool and scientific illustration. She has directed and developed the educational program *Art and Culture* with the use of art as Mediator of learning for twelve years, and she has been part of the project *The scientific drawing in the travelogue* in collaboration with the Society of Natural History of the Sea and the Center for Marine University Studies (CEMU) for three years. While maintaining her artistic activity in the field of drawing and sculpture, she participates or attends symposiums, conferences, and congresses in universities and museums, taking an active part with papers, communications, or book chapters.

# Contributors

**Miguel Molina Álvarez** Area of Pharmacology, Nutrition and Bromatology, Department of Basic Health Sciences, Unidad Asociada de I+D+i al Instituto de Química Médica (IQM-CSIC), Universidad Rey Juan Carlos, Madrid, Spain;
High Performance Experimental Pharmacology Research Group (PHARMAKOM), Universidad Rey Juan Carlos, Alcorcón, Spain;
Grupo Multidisciplinar de Investigación y Tratamiento del Dolor (i+DOL), Alcorcón, Spain

**Almudena García Carrasco** Area of Biochemistry and Molecular Biology, Department of Basic Health Sciences, Universidad Rey Juan Carlos, Alcorcón, Spain;
High Performance Research Group in the Study of Molecular Mechanisms of Glucolipotoxicity and Insulin Resistance: Implications in Obesity, Diabetes and Metabolic Syndrome (LIPOBETA), Universidad Rey Juan Carlos, Alcorcón, Spain

**Jair Antonio Tenorio Castaño** Centro de Investigación Biomédica en Red de Enfermedades Raras (CIBERER), Madrid, Spain;
Institute of Medical and Molecular Genetics, INGEMM-Idipaz, Madrid, Spain;
ITHACA, European Reference Network, Brussels, Belgium;
Madrid Technology Park (PTM), BITGENETIC, Madrid, Spain

**Patricia Corrales Cordón**  Area of Biochemistry and Molecular Biology, Department of Basic Health Sciences, Universidad Rey Juan Carlos, Alcorcón, Spain; High Performance Research Group in the Study of Molecular Mechanisms of Glucolipotoxicity and Insulin Resistance: Implications in Obesity, Diabetes and Metabolic Syndrome (LIPOBETA), Universidad Rey Juan Carlos, Alcorcón, Spain; Consolidated Research Group on Obesity and Type 2 Diabetes: Adipose Tissue Biology (BIOFAT), Universidad Rey Juan Carlos, Alcorcón, Spain

**Blanca del Carmen Migueláñez Medrán**  Area of Stomatology, Department of Nursing and Stomatology, Universidad Rey Juan Carlos, Alcorcón, Spain

**Cristina González Fernández**  External Collaborator, Area of Pharmacology, Nutrition and Bromatology, Department of Basic Health Sciences, Unidad Asociada de I+D+i al Instituto de Química Médica (IQM-CSIC), Universidad Rey Juan Carlos, Madrid, Spain

**Miguel M. Garcia**  Area of Pharmacology, Nutrition and Bromatology, Department of Basic Health Sciences, Unidad Asociada de I+D+i al Instituto de Química Médica (IQM-CSIC), Universidad Rey Juan Carlos, Madrid, Spain; High Performance Experimental Pharmacology Research Group (PHARMAKOM), Universidad Rey Juan Carlos, Alcorcón, Spain; Grupo Multidisciplinar de Investigación y Tratamiento del Dolor (i+DOL), Alcorcón, Spain

**Antonio González Ruiz**  Area of Pharmacology, Nutrition and Bromatology, Department of Basic Health Sciences, Unidad Asociada de I+D+i al Instituto de Química Médica (IQM-CSIC), Universidad Rey Juan Carlos, Madrid, Spain; High Performance Experimental Pharmacology Research Group (PHARMAKOM), Universidad Rey Juan Carlos, Alcorcón, Spain

**Elena López Guadamillas**  UCL Cancer Institute, University College London, London, UK

**Marta Martín Ruiz**  Laboratory of Molecular Biology, Hospital Universitario General de Villalba, Collado Villalba, Spain

**Gerardo Ávila Martín**  Health Integrated Area, Research Support Unit, Hospital General Universitario Nuestra Señora del Prado, Servicio de Salud de Castilla-La Mancha (SESCAM), Talavera de la Reina, Toledo, Spain

**Eva Mercado Delgado**  External Collaborator, Area of Pharmacology, Nutrition and Bromatology, Department of Basic Health Sciences, Unidad Asociada de I+D+i al Instituto de Química Médica (IQM-CSIC, Universidad Rey Juan Carlos, Madrid, Spain; Grupo Multidisciplinar de Investigación y Tratamiento del Dolor (i+DOL), Alcorcón, Spain; Pain Management Unit, Hospital Universitario Sanitas Virgen del Mar, Madrid, Spain

**Fernando Muñoz Muñoz** Clinical Diagnostic Processes and Orthoprosthetic Products, Consejería de Educación de la Comunidad de Madrid, Madrid, Spain

**Nancy Antonieta Paniagua Lora** Area of Pharmacology, Nutrition and Bromatology, Department of Basic Health Sciences, Unidad Asociada de I+D+i al Instituto de Química Médica (IQM-CSIC), Universidad Rey Juan Carlos, Madrid, Spain;
High Performance Experimental Pharmacology Research Group (PHARMAKOM), Universidad Rey Juan Carlos, Alcorcón, Spain;
Grupo Multidisciplinar de Investigación y Tratamiento del Dolor (i+DOL), Alcorcón, Spain

**Gema Vera Pasamontes** Area of Pharmacology, Nutrition and Bromatology, Department of Basic Health Sciences, Unidad Asociada de I+D+i al Instituto de Química Médica (IQM-CSIC), Universidad Rey Juan Carlos, Madrid, Spain;
Grupo Multidisciplinar de Investigación y Tratamiento del Dolor (i+DOL), Alcorcón, Spain;
High Performance Pathophysiology and Pharmacology of the Digestive System Research Group (NeuGut), Universidad Rey Juan Carlos, Alcorcón, Spain

**Patricia Puerta Gil** Instituto de Agroquímica y Tecnología de los Alimentos (IATA, CSIC), València, Spain;
Unidad Asociada de I+D+i Universitat Politècnica de València (UPV) CSIC-Centro Tecnológico de Ondas, València, Spain

**Carmen Rodríguez Rivera** Area of Pharmacology, Nutrition and Bromatology, Department of Basic Health Sciences, Unidad Asociada de I+D+i al Instituto de Química Médica (IQM-CSIC), Universidad Rey Juan Carlos, Madrid, Spain;
High Performance Experimental Pharmacology Research Group (PHARMAKOM), Universidad Rey Juan Carlos, Alcorcón, Spain;
Grupo Multidisciplinar de Investigación y Tratamiento del Dolor (i+DOL), Alcorcón, Spain

**Alejandro Salgado Flores** Thermo Fischer Scientific, Madrid, Spain;
External Collaborator, Department of Arctic and Marine Biology, UiT-The Arctic University of Norway, Tromsø, Norway

**Sergio Muñoz Sánchez** DNA Replication Group, Molecular Oncology Programme, Spanish National Cancer Research Centre (CNIO), Madrid, Spain

**Marina Martín Taboada** Area of Biochemistry and Molecular Biology, Department of Basic Health Sciences, Universidad Rey Juan Carlos, Alcorcón, Spain;
High Performance Research Group in the Study of Molecular Mechanisms of Glucolipotoxicity and Insulin Resistance: Implications in Obesity, Diabetes and Metabolic Syndrome (LIPOBETA), Universidad Rey Juan Carlos, Alcorcón, Spain

# Gaia: A Planet Dominated by Bacteria

Cristina González Fernández

Life appeared on earth approximately 3.5 billion years ago, and since then, the Earth and its living beings have been part of a structure that is capable of self-regulating as a single integrated physiological system. Gaia,[1] Mother Earth, the Goddess of classical mythology who was born out of chaos and is the mother of all, gave her name to the *Gaia Hypothesis*, formulated by the English chemist James Lovelock[2] in 1969. According to this hypothesis, the Earth behaves like a living organism that has managed to maintain the right conditions for life but, in turn, the system feeds back: the temperature of the Earth's surface, the composition of reactive gases in its inner core, the state of oxidation, acidity and alkalinity

---

[1] *Gaia* from Greek or *Gaia* from Latin is the goddess who personifies *Mother Earth.* According to Greek mythology, she emerged from nothing and, on her own, conceived *Uranus* (the Titan who personified the sky) and his brother *Pontus* (god of the sea). Later she turned her sons into husbands, and with them she became the mother of a multitude of mythological Titans and Titanides. Her Roman equivalent is *Terra.*

[2] James Lovelock is an English scientist famous for formulating the "Gaia Hypothesis": *the living matter of the Earth and its air, oceans, and surface form a complex system that can be regarded as an individual organism capable of maintaining the conditions that make life on our planet possible.* He also invented the electron capture detector, which is a very sensitive method of chemical analysis especially useful for the detection of polluting chemicals. He also experimented with freezing rodents to revive them later (cryopreservation).

C. González Fernández (✉)
External Collaborator, Area of Pharmacology, Nutrition and Bromatology, Department of Basic Health Sciences, Unidad Asociada de I+D+i al Instituto de Química Médica (IQM-CSIC), Universidad Rey Juan Carlos, Madrid, Spain
e-mail: cristina.gonf@gmail.com

M. M. Garcia (ed.), *Tales of Discovery*, https://doi.org/10.1007/978-3-031-47620-4_1

in the Earth, are maintained by the activity of life on the planet. That is, life produces and maintains its immediate environment, appearing on Earth as a planetary phenomenon. Sociologists have been able to explain this very well in the field of economics with Simmel's "The Philosophy of Money," and this could well be transferred to a biological context. Money, like life, is more than a value; it is a means of exchange that influences relations between humans (and living beings) in such a way that money begets money and life begets life.

## A World of Daisies

To deal with the criticism and try to explain how Gaia works, in 1983, Lovelock and his colleague Andrew Watson created a computer simulator they called *Daisy World*. In the first phase of the simulation, a hypothetical barren planet, with a simple atmosphere, orbits around a sun that increases its temperature in a linear mode, and causes the planet's temperature to increase as solar radiation hits it. In a second phase, the planet is planted with two different varieties of daisies: black and white. The dark color of the black daisies allows them to absorb solar radiation more easily, while the white daisies reflect it. Since the temperature of the planet is gradually increasing in line with that of the sun, in the beginning of time the planet would be cold, but over the years it would reach a temperature sufficient to be habitable. At the beginning, only the black daisies would be able to proliferate because they retain more heat than the white ones, thus dyeing the planet black. Eventually, this would lead to an increase in the temperature of the planet, regardless of the radiation received from the star. As the temperature increased, not only would black daisies be more likely to proliferate, but white daisies would also proliferate until an ideal temperature of 22.5 °C was reached, where there would be an equal number of black and white daisies. The white daisies would start growing mainly around the equator, where there is more sunlight and the temperature is higher. As solar radiation increases, the temperature on the planet rises, and the black daisies begin to absorb too much radiation, overheat, and die; they becoming restricted to the poles, which receives less sunlight, and where the temperature is lower. White daisies would then begin to predominate, keeping the temperature at habitable conditions, until finally only white daisies would remain. In the third phase of the simulation, an extreme situation would be created in which the gradual but incessant increase in temperature would make it impossible for even white daisies to take on the solar radiation, and their population would crash. Eventually, the sun's rays are programmed to decrease in power to a more comfortable level, allowing white daisies to grow which would, in turn, begin to cool the planet again. The simulation shows that from the moment life appears on the planet the temperature does not increase linearly but remains stable at approximately 22.5 °C (optimum temperature for a balanced proliferation of black and white daisies). In other words, life on the planet modifies the temperature of the environment to create more favorable conditions for itself. In new simulations, the program has been reconstructed to include other forms of life (rabbits, foxes, etc.) to see if the same effect on the temperature of the planet is produced. What Lovelock advocates always happens: life on Earth, or rather, the interactions between living beings and the environment, produce optimum conditions for their own development. Moreover, the greater the number of introduced species, the better the regulation of global temperature, proving the value of biodiversity for the planet. The Earth and all its living beings are cogs in a machine that works with precision to keep itself in balance and thus ensure its survival. If just one cog fails, it will affect the whole system, which will have to readjust and adapt to the new conditions in order to continue functioning.

## The Origin of Life: Much More Than a Single Theory

On our hypothetical planet, we can introduce new species at will, but how does it work on Earth? Imagine Earth 3.5 billion years ago: an inhospitable planet, volcanoes, an atmosphere totally incompatible with life as we know it today, ammonia, methane, carbon dioxide, lots of radiation… In these conditions it is not difficult to imagine that life originated under the Earth, protected from such conditions. In fact, apart from other religious theories such as *Genesis*, or social movements such as the *Flying Spaghetti Monster*,[3] efforts to locate life on nearby planets lacking an atmosphere compatible with life (such as Mars) are focused precisely on the study of the subsoil. If there is life on Mars, it must be there protected from the cold, from an inhospitable environment, or from extreme radiation. Understanding the past (how life originated on Earth) to understand the future (how to colonize other planets). Following this line, the theory of *Panspermia* argues that life is not a phenomenon exclusive to Earth, but is also found on other planets in the universe,

---

[3] *Genesis* is the first book of the Bible. "And God said, Let the earth bring forth grass, the herb yielding seed, and the fruit tree yielding fruit after his kind, whose seed is in itself, upon the earth: and it was so. And the earth brought forth grass, and herb yielding seed after his kind, and the tree yielding fruit, whose seed was in itself, after his kind: and God saw that it was good." (Genesis 1:11–12). "And God said, Let the waters bring forth abundantly the moving creature that hath life, and fowl that may fly above the earth in the open firmament of heaven. And God created great whales, and every living creature that moveth, which the waters brought forth abundantly, after their kind, and every winged fowl after his kind: and God saw that it was good. And God blessed them, saying, Be fruitful, and multiply, and fill the waters in the seas, and let fowl multiply in the earth." (Genesis 1:20–22). "And God said, Let the earth bring forth the living creature after his kind, cattle, and creeping thing, and beast of the earth after his kind: and it was so. And God made the beast of the earth after his kind, and cattle after their kind, and every thing that creepeth upon the earth after his kind: and God saw that it was good." (Genesis 1:24–25). For its part, the religion of the *Flying Spaghetti Monster* or *Pastafarianism* is a social movement that emerged in the United States as a social protest to denounce and oppose the teaching, in public schools, of Christian creationism and intelligent design as an alternative to the theory of evolution. Bobby Henderson (a physics graduate from Oregon State University) sent a letter to the Kansas State Board of Education in which he expressed his faith in a god resembling a huge spaghetti ball with flying meatballs (*Flying Spaghetti Monster*) who created the universe after drinking too much and, because of the monster's drunkenness, the world is imperfect.

and is *seeded* by travelling through space on board meteorites which, when they fall on the right planet, cause life to germinate on that new planet. In fact, there are microorganisms capable of withstanding the conditions of life in space, such as large temperature contrasts or high levels of radiation. Hypothetically, if there had been life on another planet before Earth, a meteorite that had hit its surface could have launched organic material towards our planet. In reality, this hypothesis does not solve the problem of the origin of life, it only changes its location. There are many different theories about the origin of life on Earth's surface. Some examples are as follows:

- The *hydrothermal source theory*, which proposes that life originated in hydrothermal vents under the sea from inorganic materials (carbon monoxide, sulfides, etc.) at high temperatures and pressure. The first proteins could have formed under these conditions.
- *Glacial theory* postulates that 3700 million years ago, the Earth was covered with ice and that this ice acted as a protective layer for organic compounds.
- The *RNA world hypothesis* argues that RNA is responsible for the emergence of life on the planet. Experiments have been done with some clays that suggest that they could have helped the formation of RNA and fatty vesicles and that, together, they would have been able to form primitive cells.
- The *theory of simple principles*, which argues that the first organic molecules must have been simple—not molecules as complex as RNA.

However, these are all theories—3500 million years ago there was no one on Earth who could document what truly happened. Any of these theories could be valid since no theory is absolutely true, but at best, not disproved. How life appeared on Earth is an unresolved question. It may have arisen spontaneously, or it may have been "seeded" on the planet. Whatever the case, we know that the first forms of life that inhabited the planet were very simple and primitive unicellular microorganisms, bacteria and archaea, which even today continue to share the planet with us. How then did life reach the level of complexity that we know today?

The *endosymbiont* or *serial endosymbiosis theory* is quite well known and accepted. Proposed by Lynn Margulis as the origin of the eukaryotic cell, this theory describes how primitive microorganisms form symbiotic associations among themselves in a specific order, giving rise to more complex eukaryotic cells. The eukaryotic cell would be the result of those primitive microorganisms engulfing each other without digesting each other. Certain ingested bacteria would have been trapped inside the cell bodies of other cells, becoming organelles and resulting in more complex eukaryotic cells. Let us go back to Earth 3.5 billion years ago, that cruel and inhospitable planet, where the first forms of life appeared: prokaryotes (bacteria and archaea). This theory postulates that, first, a fermenting archaeobacterium (a type of bacterium that lives in warm, sulfur-rich water) fused with a swimming bacterium, giving rise to the last eukaryotic common ancestor LECA (Last Eukaryotic Common Ancestor). This one would be anaerobic, so it would live in muds and sludges, puddles, ponds, rock crevices, etc.—places where oxygen is scarce. Over time, the archaeobacteria would form the current cell cytoplasm, while the swimming bacteria would give rise to cell appendages such as cilia, flagella, or sensory protrusions. Later, in a second fusion, aerobic bacteria would be incorporated, which would be the origin of mitochondria. Finally, in a third fusion, green photosynthetic bacteria would be incorporated, thus generating the chloroplasts present in plant cells (Fig. 1).

However, Margulis did not stop there; she also developed the theory of *symbiogenesis*. Contrary to Darwin's theory, which argues that evolution occurs thanks to the appearance of modifications in the offspring (random mutations) followed by natural selection by competition, the theory of symbiogenesis argues that collaboration between species has been more decisive in biological evolution than competition. That is, it is symbiotic associations and not random mutations that drove evolution. In this regard, the term *symbiosis* refers to an association between two organisms in which both receive a mutual benefit; an example of this is lichens, which are organisms that arise from the association between a fungus and an alga. Within lichens, there are some that form a very simple association and others that form more complex structures. Corals, for example, are colonial animals that live in symbiosis with photosynthetic algae from which they obtain a large part of their energy. The Portuguese Caravel is a colonial organism in which each individual in the colony performs a specialized function (some are in charge of floating, others of hunting, others of digestion, others of reproduction, etc.). Without the resident symbiotic bacteria in their digestive tracts, herbivores would not be able to obtain energy by feeding on plants. Where is the limit? When does a colony cease to be a colony and become an individual with different specialized tissues? At what point do we consider that an organism and its symbiont, essential for its survival, become a single individual? The limit is usually at the level of communication between its cells. We humans like to classify things, put each one in a well-differentiated drawer, and give each drawer a name, but sometimes nature is not so organized. It does not classify or set the limits, rather it evolves and optimizes. When, in these associations, an organism lives inside another we speak of *endosymbiosis*, an example of which is bacteria that live inside some species of insects. These associations can become so close that the insect cannot survive without these bacteria. However, when we talk about *symbiogenesis*, we are taking it a step further. The association between host and symbiont becomes so close that the symbiont transfers part or all of its genome to the host, resulting in the appearance of new morphologies, tissues, metabolic pathways, behaviors, organs, organ systems, or other evolutionary novelties that make it better adapted to the environment than its individual components.

We could say that eukaryotic organisms present communities of prokaryotes that have coevolved and are closely integrated. We humans would be the result of a set of millions of highly specialized and intercommunicated bacteria collaborating in order to survive; bacteria are the forms of life that dominate the planet. Wherever there are plants there will be their chloroplasts obtaining energy from the sun and wherever there is an animal there will be their mitochondria providing energy. No individual (be it protist, fungus, animal, or plant) can survive without

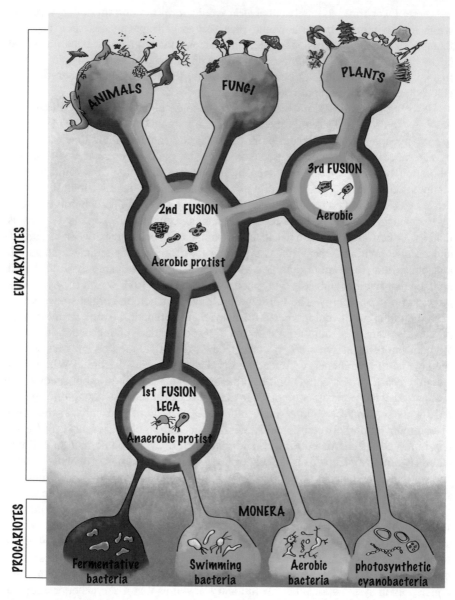

**Fig. 1** **Serial endosymbiosis theory**. The most accepted theory postulates that the organisms that populate the Earth today are the result of fusions and associations of more elementary organisms. Through comparative studies, vestiges of these more primordial organisms can still be observed by analyzing the cells that compose us

the existence of other individuals, not even the human being. Only bacteria are able to survive without the existence of other individuals. In her book "Symbiotic Planet," Lynn Margulis tells how her son Zach asked her one day: "Mom, what does the idea of Gaia have to do with your symbiotic theory?" And to answer this question she used a quip made by one of her students, who said that "Gaia is simply symbiosis as seen from space." Lynn defined symbiosis as "the living together and in physical contact of organisms of different species." In Gaia, all organisms remain in contact and share the same atmosphere, oceans, rivers, and seas. As Lynn Margulis said, "we are still symbiotic beings on a symbiotic planet."

If we consider that Gaia is a single integrated physiological system capable of self-regulation, it is also possible that this balance could be lost, as it can be lost in other physiological systems such as in our own organism, thus giving rise to disease. Could Gaia then become ill? In recent years, the signs of illness that our planet is showing have been evident. Global warming and the resulting climate change bring with it serious consequences for the Earth's flora and fauna (including humans). Some of these consequences are melting glaciers and ice caps with the subsequent rise in sea level, extreme weather events, extinction of species, desertification, mass migrations. This phenomenon is already underway, and it is important to assume that climate change cannot be avoided. What we can do is try to reduce its effects, and adapt to the change. Many experts date the beginning of this great change to the Industrial Revolution (eighteenth-nineteenth century). Since then, population growth has increased significantly, with the consequent excessive increase in the consumption of resources, including energy obtained mainly from fossil fuels, the increase in the emission of greenhouse gases into the atmosphere, deforestation, the destruction of marine ecosystems, and so on. Such has been the impact of human activity on the planet that there is already talk of a new geological era: the *Anthropocene*.

The Earth has already suffered at least five great global extinctions in the past. Humans tend to think that we are the pinnacle of evolution, the "great masterpiece" of nature, and yet we do not know how to live in symbiosis with everything around us; instead, we behave more like a virus. Faced with this new landscape, many species will become extinct, others will survive (hopefully including ours), and symbiosis will undoubtedly play an important role in the adaptation of species to the new environment. As has already happened on other occasions, the system will change, and the conditions will be different—but Gaia will survive.

## Bibliography

Chapman MJ, Margulis L (1998) Morphogenesis by symbiogenesis. Int Microbio 1(4):319–326

Delisle K (2009) Cells and evolution. In: Margulis L, Dolan MF, Càtedra de Divulgació de la Ciència. The beginnings of life. Evolution on the Precambrian Earth. Publicacions de la Universitat de València, Valencia, p 23

Guerrero R, Margulis L, Berlanga M (2013) Symbiogenesis: the holobiont as a unit of evolution. Int Microbiol 16(3):133–143

Lovelock JE (1985) Gaia, a new vision of life on Earth. Ediciones Orbis, Barcelona

Margulis L (2001) The conscious cell. Ann N Y Acad Sci 929:55–70
Margulis L (2002) Symbiotic planet. A new point of view on evolution. Editorial Debate, Barcelona
Watson AJ, Lovelock JE (1983) Biological homeostasis of the global environment: the parable of Daisyworld. Tellus 35B(4):286–289

**Cristina González Fernández** holds a degree in Biology from Universidad de Alcalá de Henares (UAH) and a Ph.D. from the Doctoral Program in Pain Research at Universidad Rey Juan Carlos (URJC). She has developed her professional career as a professor and researcher in the Area of Pharmacology, Nutrition and Bromatology of the Department of Basic Health Sciences of Universidad Rey Juan Carlos, investigating the effects of drugs, nutraceuticals and egg white hydrolysates on the cardiovascular system in animal models. She has participated in outreach activities (e.g., *Science Week*, the *European Researchers' Night*, the *QUO Gala of science and innovation*) and in science and gastronomy conferences.

# Me, Myself and My Microbiota

Patricia Puerta Gil [ID]

Bacteria—what comes to mind with this word? In general, when we hear words such as bacteria, microbes, or microorganisms, we tend to think of pathogens, dirt, diseases, infections, epidemics, or deaths. Why have we arrived at such a dismal association of these words? No doubt we have skewed the concept of the microbial world, causing our subconscious to give it this negative nuance almost without realizing it, and unleashing a fear (sometimes irrational) of these great unknowns invisible to the human eye. Although they have earned a bad reputation, in this chapter, we will try to dismantle this belief, shift the paradigm, and provide objective data that corroborate that most of the microorganisms are not responsible for diseases; on the contrary, they can be used to benefit our own survival.

P. Puerta Gil (✉)
Instituto de Agroquímica y Tecnología de los Alimentos (IATA, CSIC), València, Spain
e-mail: ppuerta@iata.csic.es

Unidad Asociada de I+D+i Universitat Politècnica de València (UPV) CSIC-Centro Tecnológico de Ondas, València, Spain

© The Author(s), under exclusive license to Springer Nature Switzerland AG 2024
M. M. Garcia (ed.), *Tales of Discovery*, https://doi.org/10.1007/978-3-031-47620-4_2

## The Origin of the Eukaryotic Cell: Evolution by Bartering?

Most of the living things around us are microscopic and intimately related to the cycle of life. And no, this is not an homage to the *Lion King*. I am talking about the energy-matter transformations[1] that occur cyclically in a constant balance within the ecosystem. Microbes (or microorganisms) were the first inhabitants of our planet; they live comfortably in other environments where others have not been able to adapt, and may be the only form of life on other worlds. Since those first associations between molecules in the pools of a primordial Earth that gave rise to a "genetic protomaterial" precursor of DNA,[2] microorganisms have managed to obtain energy and to self-perpetuate. They are our ancestors. They occupy a prominent place on the tree of life, and from them all the other forms of life we know, including human beings, have been formed. However, they are not just a starting point; we have coevolved with them and they have made us what we are today. If animal and plant life were to magically disappear from the Earth, you would still be able to see our silhouettes and internal organs, the surface of the ground, and the trees because they are all covered in bacteria. Think of a Christmas tree: even if the tree itself could not be seen in the dark of night, its lights would still define its outline.

There is increasing evidence of the intimate associations we maintain with microorganisms. These connections have been a constant throughout our evolution and are based on rudimentary rules of coexistence. One such example of an interaction that favors and improves the survival of both organisms: the host and the bacterium. One of the most striking examples of these early relationships is the mitochondria. These organelles housed inside our cells are able to provide the main source of energy in a eukaryotic cell. Currently, the most accepted theory to explain the origin of our cells is the endosymbiotic theory proposed by Lynn

---

[1] According to the Laws of Thermodynamics, energy is neither created nor destroyed, it is only transformed. The energy that comes from the sun's rays and hits the leaves of a plant is absorbed and transmitted from one molecule to another until it generates a reaction that transforms water and $CO_2$ into sugars and oxygen. The plant uses these sugars for its growth and, when it dies, the components of the plant are decomposed, degraded into simpler particles that serve as nutrients (energy supply) to other organisms. Thus, energy continues to be transmitted and the cycle continues.

[2] The most accepted theory about the origin of life on Earth was proposed by the Russian biologist Aleksandr Oparin in 1924, and focuses on the chemical evolution of inorganic to organic molecules in reducing environmental conditions; that is, in the absence of oxygen. In that primitive Earth would originate, thanks to electrical energy and ultraviolet radiation, the first organic molecules rich in carbon, nitrogen and hydrogen, precursors of proteins, nucleic acids, and other biomolecules, known as the primordial soup. These first biomolecules would interact with each other in the soup and, in particular, the precursors of the genetic material would be enclosed in mycelial structures that would protect it from the environment. In 1953, the scientist Stanley Miller proposed an experiment in his laboratory to corroborate Oparin's theory. Recreating the supposed conditions of the primitive Earth, he managed to successfully produce organic compounds from methane, ammonia, water, and nitrogen, demonstrating that biological polymers can be formed from simple molecules with a suitable energy source.

Margulis in the 1960s. According to this, mitochondria evolved from a prokaryotic ancestor, capable of metabolizing oxygen, that was phagocytosed by an ancestral proto-eukaryotic cell, which was still anaerobic and unable to live in the presence of oxygen (Fig. 1). In this way, a symbiotic relationship was established between the two cells, giving the eukaryotic cell the ability to breathe aerobically and use ambient oxygen to increase its ability to obtain energy. This also managed to modify the concentration of gases in the atmosphere, increasing the amount of oxygen. The organism itself, by the mere fact of existing, generated changes in its immediate environment. In this phagocytosis, there was also an exchange of genetic material (transfer processes).

Throughout the evolutionary history of the planet, organisms have been gaining complexity: unicellular organisms consisting of a single cell aggregated at some point and started exhibiting behaviors akin to multicellular organisms, with clusters of cells specialized for different tasks. This is how it works in humans. As multicellular organisms, we have organs and systems specialized for exclusive functions. Associations with prokaryotic microorganisms have also become part of our body systems, producing useful molecules that are now sometimes vital to us; they can also be involved in other tasks, such as regulation of the immune system.

## Cellular Associations. Unity Is Strength

As the most elementary forms of animal life diversified, the interaction with microbial cells continued to shape the way we evolved, influencing the development of the digestive, respiratory, or urogenital system. We have grown with them as a species and integrated them throughout evolution. The shape and structures of the tissues we have reflect the pressure the environment has exerted on natural selection. That is, when we came out of the oceans, a life on land with little interaction with water meant that we were subjected to long periods of dryness and abrasions on the entire surface of our organism. Only those organisms with a more protective and impermeable skin layer were able to adapt efficiently to the new environment. The colonization of our body surface by certain groups of microorganisms also served as a barrier and as an additional nutritional tool for the cells that covered our skin. The microbial distribution on our organism reflects the adaptations that took place when we went from marine life to the colonization of the terrestrial environment. By this, I mean not only the bacteria present on our skin, but also those located throughout our respiratory system and digestive tract. In fact, the microbial flora we can find in the intestines of a fish and a human is completely different, basically because the environments are so different.

Since those times, there has been such a strong functional and ecological relationship between the communities that now we could not live without them. For example, our microbial flora is essential to the absorption of fatty acids, calcium, magnesium, and other nutrients. Without it, the body would not be able to assimilate nutrients from our diet, and they would cross directly from the intestine to

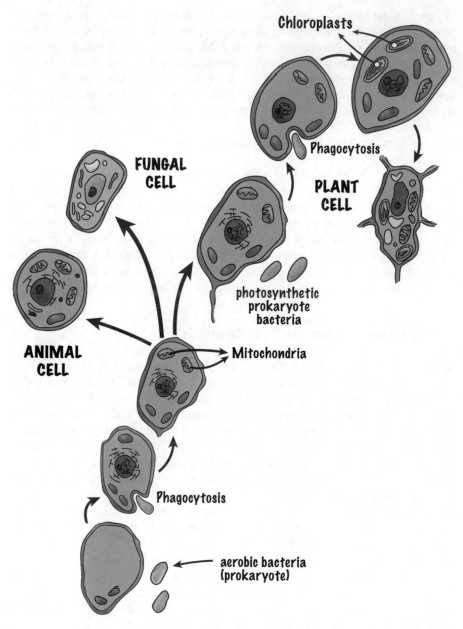

**Fig. 1** **The endosymbiotic theory proposed by Lynn Margulis**. Schematic view of the original theory postulated by Lynn Margulis on the origin of the different eukaryotic cells

the anus. These microorganisms also contribute to immune functions and intestinal motility, help synthesize certain vitamins, and break down complex molecules into nutrients that are easier to assimilate. Our existence depends critically on the presence of these bacterial partners that we have adopted as our own. This very favorable cooperation between microbial cells and our own, known as *symbiosis* or *mutualism*, is made possible by the establishment of a body-wide balance between them and us. Although mutualism is the predominant type of relationship, we also maintain some commensal relations[3]: we provide them with shelter and food, but in return we receive no benefit; we remain rather indifferent. It seems that as long as they do not harm us, things can stay like that, and there is no need to fight. This whole set of communities that live in balance with us is what is known as the body's normal microbiota or microbiome, referring to the entire habitat of microorganisms, their genes, and environmental conditions.

Every inch of the surface of our body, with its different characteristics—temperature, humidity, pH, availability of nutrients, etc.—constitutes an ecological niche[4] to which different complex microbial ecosystems have adapted and specialized. In such a wide variety of scenarios, each specific group of microorganisms, having different aptitudes for tissues to colonize and proliferate, has found a place to settle. In healthy individuals, the normal microbiota is found lining the mucous membranes of the body, specifically the gastrointestinal, respiratory, urinary and genital tracts, as well as the skin, external ear, ocular conjunctiva and mouth. It may not sound like much, but in reality, these surfaces constitute one of the most densely populated microbial ecosystems on the planet.

Since we have this close relationship with our microbial inhabitants, let us get to know each other a little better. Most of the microbiota in general is made up of bacterial cells, but we also have some types of fungi, protozoa, archaea and viruses[5] living in our body.

---

[3] When two species or communities interact in an ecosystem sharing activities or requirements for their vital development, it is possible for both to benefit, harm, or not affect each other. This intimate association is called symbiosis, and it can take three forms: mutualism, commensalism, and parasitism. In the case of mutualism, both species benefit; in commensalism, one species benefits from the interaction without harming the other; in parasitism, one species benefits to the detriment of the other, which it harms.

[4] In ecology, a niche refers to the relational or functional position of a species or population in an ecosystem. It is a broad concept, and is not only related to the physical space (which would be the habitat) but also to the functional role of an organism in the community, and its position within the environmental variants, such as temperature, humidity, or pH, i.e., how a species acts under certain habitat conditions and under the influence of other species.

[5] Bacteria and archaea belong to the group of organisms known as prokaryotes, which are fundamentally characterized by having their genetic material dispersed in the cytoplasm, without a defined cell nucleus. In addition, they are usually unicellular, lack mitochondria, and are surrounded by a plasma membrane composed of peptidoglycans. In contrast, eukaryotes (animal, protozoan, plant, and fungal cells) have organized genetic material in a nucleus, are larger, are

Determining the exact number of microorganisms we harbor has not been an easy task, but it is estimated that $10^{14}$ bacteria cover our body; that is, 100 trillion! We have 10 times more bacteria than human cells! Nine out of ten cells in our body are not truly ours. You might as well say that we are more of a prokaryotic organism than a eukaryotic one, right? Most of the cells in the body are found in the blood, so bacteria populate the rest of the body's surfaces. As I mentioned earlier, if our body were to disappear, and we could illuminate the bacteria that cohabit it, they would look like the lights on a Christmas tree, marking its perimeter and nooks and crannies. Moreover, these 100 trillion bacteria correspond to 1000 different species. Overall, each one of us is a walking ecosystem with very rich biological diversity—a mega-community known as a *holobiont*. It is suspected that we may be harboring a multitude of other species, but the difficulty in isolating many of these classes makes it impossible to characterize some other individuals.

Through the application of new molecular techniques, such as the sequencing of genomic material, it has been estimated that the number of microbial genes is 100 times greater than that of human genes. Thanks to these techniques, it has been possible to advance the taxonomic knowledge of such vast diversity, and undoubtedly, in the future we will learn more and more about our guests, and ourselves with them. We said before that many of the cells of the body are found in the blood, but what about bacteria? Where are they found? Let us see how they are distributed in the most inhabited areas of the body.

## Skin Surface

The skin covering our body is our largest organ, with an average surface area of approximately 1.8 $m^2$. Each area has unique conditions, providing specific microhabitats for the microbiota. Depending on the temperature, pH, humidity, available nutrients, fat, or solar radiation to which each area is subjected, we find a different profile of microorganisms. Although the proportion, quantity, and identity of the microbiota is distinctive to each person, in general, each square centimeter of our skin is home to more than 10 million bacteria and fungi.

At least 150 different species colonize the surface of the skin, the grooves, accesses or ducts of hair follicles, and sweat and sebaceous glands, but we predominantly found bacteria belonging to the genera *Corynebacterium*, *Propionibacterium*, and *Staphylococcus*. They form a superorganism that is in constant communication with our skin, acting as a protective barrier (mechanical and chemical) against possible invaders, and determining the permeability between the internal and external environment. The great diversity and balance present in these niches are key to the maintenance of healthy and resistant skin. For example, the hormonal alterations that take place during puberty are responsible for the

---

usually multicellularly associated, and their membranes are composed of phospholipids. However, both prokaryotes and eukaryotes are cellular life forms, unlike viruses, which are considered acellular and need another cell which they parasitize in order to complete their life cycle.

increased sebaceous secretion in areas such as the face and back. For *Propioni-bacterium acnes*, this sebum-free "buffet" is a feast, and proliferates out of all proportion, causing uncomfortable acne.


## External Ear

In reality, the areas of the external auditory canal and ears are an extension of the cutaneous microbiota, so the representation of species present in the ear micro-biota is similar to that of the skin. The situation here, however, is somewhat delicate, as the presence of some *Staphylococcus* and other commensals or poten-tial pathogens, such as *Streptococcus pneumoniae*, could cause infection in areas of the middle ear if they were to penetrate the tympanic membrane.


## Ocular Conjunctiva

The ocular microbiota also resembles that of the skin for the most part, although the lysozyme present in tears limits bacterial growth to avoid possible infections. However, this area is susceptible to certain infections, such as those that cause conjunctivitis, which can be both bacterial and viral.


## Respiratory System

Most of the microorganisms present in the respiratory tract are not usually pathogenic, and are rarely associated with disease (as long as they do not invade normally sterile areas). Among the most common bacteria are genera with dif-ferent oxygen tolerances belonging to the phyla *Bacteroidetes, Firmicutes,* and *Proteobacteria.* In the upper respiratory tract, more than 300 different species have been described, while in the lower respiratory tract, the stable microbiota is similar but with less density and diversity.


## Genitourinary System

The high relative humidity of this area means that there is a wide range of between 20 and 500 bacterial species in sections of the outermost tract, whereas the internal urinary tract, as well as the bladder, is normally devoid of microbiota.

In the case of women, the vaginal mucosa is home to its own microbiota that protects it from undesirable microorganisms. The population in this area is highly heterogeneous and is influenced by various hormonal factors, which vary through-out life and throughout the menstrual cycle. More than 250 different species can be found, although bacteria belonging to the genus *Lactobacillus* have a clear pre-dominance. When these populations are diminished by physiological or external disturbances, infections caused by opportunistic pathogens, such as those caused by species of the fungus *Candida,* can occur. The absence of these lactobacilli and their consequent pathologies in this peculiar ecosystem highlights once again the importance of their presence in the defense against harmful infectious agents.

## Gastrointestinal Tract

Finally, we explore the gastrointestinal tract. The human digestive system is colonized by microorganisms from the oral cavity to the rectum, reaching a maximum in the large intestine of up to $10^{12}$ bacterial cells per gram, making it the largest microbial ecosystem in the body, with rich and active populations. Some consider it a truly essential organ integrated into our biology, interacting with the other organs and systems of our body as one more.

The intestinal microbiota, formerly known as the intestinal flora, consists mainly of bacteria and, to a lesser extent, archaea, although some fungi and protozoa may inhabit the intestinal flora as innocuous commensals. The categorization of diversity in this niche has been a complex task, as it is particularly difficult to recreate the conditions of the digestive tract ex vivo. The most modern molecular techniques have been able to determine that in the intestinal tract of an adult person we can find approximately 500 to 1000 different species, although more recent studies suggest that we could harbor many, many more, even between 15,000 and up to 36,000! In addition, it is important to keep in mind that there are many bacteria that are allochthonous; that is, they appear transiently in the intestinal microbiota, unlike the autochthonous ones that are permanent residents.

Each person and the specific circumstances in which they are involved are a world of their own, and their microbiota are no different. Each of us harbors radically different collections of microbes. This microbial profile can vary between individuals not only in diversity but also in the relative abundance of each type. Leaving aside the particularities of each one, there are two major groups that we all share and that represent almost 85% of the entire microbiota. These are the groups known as *Bacteroidetes* and *Firmicutes,* and they have the greatest diversity. However, there are some genera, such as *Bifidobacterium,*[6] that, although they represent only 5% of the microbiota community, have great importance in the functioning and global dynamics of the ecosystem. In addition, there are certain

---

[6] Bacterial phylogeny groups the different families of bacteria according to molecular genetic studies, and composition of the surrounding wall. *Bacteroidetes* are a phylum of Gram-negative bacteria, i.e., their cell wall does not stain dark blue or violet by the Gram staining method, and are mainly anaerobic. *Firmictues* (and within them, the genus *Bifidobacterium*), on the other hand, have mostly a Gram-positive cell structure. Gram staining responds to the chemical characteristics of the cell envelope structure. Gram-positive bacteria have a thick peptidoglycan cell wall surrounding the cytoplasmic membrane, which gives it great resistance. It is responsible for the dark blue or violet Gram staining. In contrast, the peptidoglycan wall of Gram-negative bacteria is thin and has an additional outer membrane. In the Gram stain, they do not stain dark blue, but pink.

types of bacteria that are always present in this niche, but with different relative abundances depending on the area.

The microbial community is structured in the intestine based on endogenous factors of the host, such as its genotype, gastric and biliary secretions, peristaltic movements, or food intervals. Although undoubtedly, what most influences the composition of the communities that inhabit us are exogenous factors, such as the type of diet, the structure and viscosity of the food, or the intake of microorganisms and drugs, especially antibiotics. However, how is each community distributed throughout this rich gastrointestinal ecosystem? As we have already learned, that depends on the conditions of each compartment.

The journey begins in the mouth. In the oral cavity, there are numerous microenvironments conditioned both by the oral anatomy itself (teeth, cavities, gums) and by contact with food or saliva; it therefore has a particularly abundant and diverse microbiota. In a healthy mouth, the microbiota forms a thick layer, a biofilm adhered to the teeth that protects us from the acid that comes from food processing and from the pathogens that we ingest. However, disorders such as gingivitis or cavities are becoming increasingly common in the modern era. Our most distant ancestors were not overly concerned with oral hygiene and possessed decay-free, relatively healthy and white teeth, as evidenced by studies of calcified dental plaque in prehistoric dentures. What has changed with evolution? What marked a before and after in the emergence of cariogenic species was the era of the *Industrial Revolution*. It was the largest change in the production and technological processing of food since prehistoric times. Suddenly, large quantities of refined flours and sugars become generously available, and from that moment on, the wide diversity of bacteria we had in our mouths disappears. In most of the dentures studied, another dominant class of bacteria associated with gingivitis or caries appears thereafter. They proliferate easily with our modern diet, being able to ferment the sugars in sweet treats, and producing demineralization of tooth enamel as a result of a more acidic pH. Changes in eating habits, therefore, seem to be a threat to the balance of our mouth; beneficial bacteria do not tolerate these new conditions of high acidity so well, favoring instead other more harmful species. Undoubtedly, something to consider when choosing dessert.

Continuing down the esophagus, we come to the stomach. Before the 1970s, it was not thought that there could be a stable community in the extremely acidic conditions of our stomach. That is as far as the vast majority of the microorganisms we ingest go; they do not usually survive beyond this chemical filter. However, communities capable of surviving and being present in this environment have been identified, as in the case of the well-known *Helicobacter pylori*. This bacterium synthesizes ammonia in its immediate environment, reducing acidity and allowing it to colonize the gastric mucosa for years or even for life. More than 50% of the world's population is a carrier of *H. pylori,* and although in most cases it is an asymptomatic condition, the presence of this bacterium is associated with an increased risk of ulcers, gastric cancer or other pathologies, but only in a part of the population. If it has coevolved with us in a more or less controlled way, and has colonized us for more than 50,000 years, how is it that it causes disease in some cases? The answer lies in the existence of multiple strains of this bacterium that colonize us, which are responsible for different levels of virulence. Some studies suggest that the coexistence and competition between different strains keeps the pathogenicity of some of them at bay. Parallel to socioeconomic development and improved hygienic conditions in the recent era, the transmission of this commensal microbe has decreased, along with the average number of strains that colonizes a person. Therefore, if a more virulent strain becomes dominant, the balance between communities is upset; that is when the problems increasingly known among the population can appear.

Leaving the stomach behind and moving on to the small intestine, things start to change. Acidity levels drop, reaching pH values close to neutral. This section is the transition zone to the large intestine, where the greatest diversity and population density is reached. In the terminal part of the small intestine, the available oxygen is reduced, and the communities that dominate from here onwards are anaerobic, i.e., they manage quite well without air. Already in the large intestine, the bacteria undergo an increase in their population to as much as 1 billion per milliliter in the terminal part of the colon.

Why are all these microbial tenants that we harbor throughout the digestive system so important? As we have been emphasizing from the beginning, they are essential for the performance of some vital functions, without which our survival would be seriously jeopardized. The microorganism community in the gastrointestinal tract performs numerous essential functions in a variety of ways:

- Protective function: They help prevent the invasion and colonization of opportunistic species by exerting a barrier effect against them, and by synthesizing antimicrobial substances that prevent their growth.
- Function on the development and maturation of the immune system associated with the gastrointestinal tract: The microbiota is able to stimulate the regeneration of the intestinal epithelium and influence cell differentiation. Thus, the response against pathogen invasion is highly effective in combination with the above protective function. In addition, some bacteria release compounds that avert inflammation, preventing the immune system from overreacting when it is not necessary.
- Regulatory function of the nervous system: The gut microbiota and the central nervous system are bidirectionally connected through the vagus nerve. This brain-gut axis integrates neuronal, hormonal, and immunological signaling systems between the gut and the brain. Many of the bacteria we harbor are capable of synthesizing some important brain neurotransmitters. Such is the importance of the nervous tissue surrounding the gut that some scientists consider it to be our "second brain."
- Metabolic and nutritional function: Thanks to the great enzymatic diversity and metabolic pathways that exist between them, microorganisms are able to process components from the diet that we cannot digest ourselves, such as dietary fiber (known as *prebiotics*). Microbiota ferments synthesize valuable metabolites and essential nutrients, such as vitamins B and K. The products resulting

from this microbial metabolism can be utilized even by other microorganisms, and provide up to 10% of our energy requirements. In addition, they can degrade toxic substances from the diet, aid in the absorption of certain electrolytes and minerals, and promote intestinal motility.

The effectiveness with which the energy intake from our diet is managed also depends directly on our microbial profile. In recent years, it has become clear that the predisposition to obesity has much to do with our microbial component, as it exerts a great influence on complex lipid metabolism. Through various investigations, it has been possible to associate the relative abundance of certain bacterial populations in the intestinal tract with the increase in body mass. Thus, depending on the bacteria we have in greater or lesser proportions, it is possible that we metabolize food in a very different way. In particular, the balance between *Firmicutes* and *Bacteroidetes* has a fundamental impact and can be considered a marker of obesity. In obese individuals, a higher relative abundance of the *Firmicutes* group has been observed, associated with increased fat absorption and increased inflammatory processes. In contrast, in thin individuals, the groups that predominate in greater proportion are *Bacteroidetes*. Historically, having bacteria in our bodies to provide us with plenty of energy was considered very beneficial, as it allowed us to go for a long time without eating, and survive on more intermittent food sources. Now, however, our energy needs are quite different, and given certain unhealthy eating habits, overweight predominates in the developed world.

## The Factors that Affect Our Microbiota also Shape Its Character

There are few physiological and immunological parameters that are not profoundly affected by the presence and nature of the body's normal microbiota. The microorganisms that colonize us have an enormous influence on our body and our way of life in a fundamental, although in many respects still completely unexplained, way. What is certain is that without a proper bacterial balance, the entire microbiota community and its interactions with its host do not function well. We are continuously exposed to a multitude of factors that can lead to *dysbiosis*, an altered state of the intestinal microbial community that can be both qualitative (predominance of species different from the usual ones) and quantitative (decreased concentration of beneficial microbiota). Specifically, imbalances in the intestinal microbiota have consequences beyond the digestive system, since to achieve a state of integral health of the organism, it is necessary that our microbiota is also healthy, especially the microbiota associated with the gastrointestinal tract. When this balance is broken, different pathologies can appear, as suggested by comparative studies of the microbiota between unwell and healthy individuals. At a local level, this dysbiosis has been related to manifestations of inflammatory bowel disease (including Crohn's disease and ulcerative colitis), diarrhea caused by invasive pathogens such as *Clostridium difficile*, or even an increased risk of developing colorectal cancer. The microbiota would be involved in this process through proinflammatory

metabolic activity as a consequence of maintaining a diet rich in fats and animal proteins but poor in fiber from vegetables, fruits and cereals. At a systemic
level, the alterations have been associated with metabolic diseases such as obesity,
malnutrition or metabolic syndrome, allergies, asthma, or nervous system diseases
such as depression, anxiety, or insomnia, due to the aforementioned brain-gut connection. The consequences of such alterations can occur in a matter of days after
the ingestion of pathogens or antibiotics, or in the long-term as a result of certain
eating habits, for example. Therefore, to prevent the effects of possible imbalances,
it is necessary to understand the different factors that influence the composition and
state of our microbiota.

As we mentioned at the beginning, the composition of the human microbiome
is unique to each individual. Although a good part of this uniqueness in our microbial DNI is due to genetic factors, studies of genetically identical twins have shown
that environmental factors are an even greater influence. While it is true that the
microbiomes of genetically identical twins are more similar to each other than to
other relatives, they still differ greatly in terms of composition and structure. One
of the most influential factors is diet. What we eat has an extraordinary power to
modulate the composition and dynamics of our microbiota. When we eat, we are
not only feeding our cells but also our millions of guests, who hungrily await their
ratio. That is why it is of utmost importance that our diet is balanced and varied,
even if we have heard this recommendation ad nauseam. Beyond the most obvious effects, such as weight gain, cholesterol, or hypertension, our diet determines
the functional metabolism of the microbiome so that its activity affects our overall
health. In his 1850 work The Teaching of Nutrition, German anthropologist Feuer-
bach stated, "we are what we eat"; not only that, rather we are what our microbiota
does with what we eat. Remember, the dietary fiber that cannot be absorbed by
our organism is metabolized by our inhabitants—it is what nourishes them. As
different studies have shown, dietary habits poor in fiber and rich in sugars, proteins, and animal fat cause the progressive loss of microbial richness and diversity
and, therefore, of the metabolic and functional potential of these populations. On
the other hand, a diet rich in fruits, tubers, vegetables, and complex carbohydrates
present in unrefined cereals favors the development of a rich and diverse microbiota. In the latter scenario, the dietary fiber-fermenting species contained in these
foods would feast fully, would be cared for, well-nourished and in balance, serving
their beneficial functions.

Antibiotic intake is another highly influential external factor that directly dis-
turbs the composition and stability of the microbiota. The effect of their repeated
use on the entire microbiome is being studied to determine the exact impact,
both at the community level and on specific strain populations. Some studies have
already shown that in adults, the microbiota does not show resilience to repeated
consumption, i.e., it does not recover its previous state. In fact, there are populations of bacteria that completely change their ecological relationship with the rest
of the ecosystem, or even cease to exist. Additionally, excessive use during the
postnatal stage and infant age has a direct impact on the acquisition of microbiota,
which can have consequences of varying severity *later in life*. Such alterations in

the proper establishment of the microbiome during the first years of life affect communication with the host, its development, and long-term health. There is increasing evidence of an association between such alterations and an increased risk of developing obesity, asthma, allergies, inflammatory bowel disease, and other metabolic and immunological conditions later in life.

During the first years of life, it is not only the alterations caused by medication that are decisive but also the way in which we are born that determines the establishment of our microbiome. Vaginal birth provides the baby with the first bacteria that colonize its previously sterile organism. Passing through the birth canal, the newborn comes into contact with specific communities of microorganisms that it needs to survive on the outside. This first exposure is crucial for our health, as these are mostly lactic acid bacteria that will be needed for the digestion of milk. In addition, the first colonizers are responsible for training the immune system. They tell it which are the good ones and, in this way, the defenses learn to distinguish between beneficial or harmless microorganisms and pathogens. They are one of our first defensive barriers when we come into contact with the outside world. If the wrong microorganisms colonize first, the balance could be upset from the start; the immune system would have a harder time learning to distinguish the good from the bad and might attack the wrong microorganisms, paying the consequences later. We will not be able to exert an effective immune response if we do not have a balance in our microbial ecosystem. This seems to be the basis of certain autoimmune diseases where antigens from the gut microbiota represent a large enough stimulus to trigger an inflammatory response.

In Caesarean deliveries, on the other hand, the first microorganisms that come into contact with the newborn are not the lactic acid bacteria of the birth canal but epithelial bacteria from both the mother and the clinical staff and other environmental bacteria. This mode of birth does not directly imply that the baby will fall ill, but many studies associate it with a greater predisposition to develop certain diseases later, such as asthma, allergies, or other immune disorders that the beneficial microbiota can help prevent. Caesarean sections are undoubtedly essential in some cases, so it is extremely important to select the use of this technique carefully. In recent years, some interesting solutions have been proposed to incorporate the beneficial effects of vaginal delivery when it is necessary to resort to caesarean section. One example would be to introduce gauze with physiological saline into the birth canal for approximately an hour to impregnate it with the protective bacteria before performing the caesarean section. Immediately after delivery, gauze was used to wipe the newborn's skin, mouth, ears, or eyes. It is still too early to be sure that this baptism of microbes can reduce the risk of future illnesses, but incorporating this simple practice in cesarean births could be of great help. In this way, the immune system would be better trained to efficiently defend against the right microorganisms.

The lifestyle we lead also plays an important role in our microbial composition. It is significantly influenced by where we live (whether in the countryside, suburbs, or city) by physical interactions with other people, exposure to different occupational environments, the regular contact we have with the outside environment or even with animals. It has been shown that living with pets is associated with a richer and more diverse microbiome composition, which is ultimately extremely beneficial for the health of our microbial ecosystem. Exercise also appears to modulate the structure of the microbiota in a way that reduces certain inflammatory states. Conversely, lack of sleep or stress causes detrimental changes in their balance, negatively affecting certain microbial communities and consequently our health.

## Accompanied by the Microbial Universe, in Health and Disease

Thus far, we have seen in a general way where our microscopic companions live, the beneficial functions they perform, and the importance of maintaining a balance between all of them. However, we may still have some doubts: Are they always harmless? Where is the fine line between homeostatic balance and disease? Normal microbiota need to be contained in their specific territory, and have restricted access to other areas. When a certain group of microorganisms invades other unusual niches where they are not welcome, problems associated with infections appear. An example of alteration by invasion of another niche by the wrong microbe is found in cases of cystitis. The most common form of this infection occurs when the bacterium *Escherichia coli*, typical of the normal microbiota of the intestine, invades the urethra and colonizes the urinary tract and can even reach the kidneys. In this case, our beloved *E. coli becomes* an opportunistic pathogen when it colonizes a niche other than its own, such as the urinary system.

On the other hand, there are certain areas in the human body that remain sterile under normal conditions, where no evidence of any type of microbiota is found. In fact, colonization of one of these sites usually leads to the onset of infection and disease if the immune system is not able to cope with the invasion. Some of these sectors with prohibited access are the central nervous system, the blood, the pericardium, the inner ear, the bronchi, the kidneys, or the uterus. This is how the dark side of the microorganism is revealed. When it invades and colonizes another organism, harming and endangering the organism's normal functions—and even its survival—for its own benefit, it becomes a *parasitic* relationship. It is no longer part of the normal microbiota and is no longer welcome; it is then an opportunistic disease-causing pathogenic microorganism.

Different microorganisms are restricted to specific niches within the body that provide them the most advantageous conditions in which to settle; however, microorganisms are also limited to specific hosts where they can readily proliferate. As we have seen, spreading afield of their niche may have dramatic consequences, as does cross-species transmission. Moreover, since colonization may occur at ease due to lack of competence and no previous exposure, host jumps typically imply increased virulence, and so the ability to cause damage in the new hosts. During a particular evolutionary period, the human species suffered these effects more markedly than at other times in history (although it is a phenomenon that occurs with some frequency in nature). Until the Neolithic period, humans were basically dedicated to hunting and gathering for survival, and populations were rather sparse and scattered. However, the emergence of agriculture and animal husbandry approximately 10,000 years ago made life easier, and brought about the *Neolithic Revolution*. It allowed food resources to become much more abundant and within reach, causing us to grow as a species in an extraordinary way. We went from leading a nomadic lifestyle to settling in concentrated populations, which made us attractive hosts for lurking pathogenic species that, until then, were part of the normal microbiota of other species. These early population clusters in limited areas provided the perfect setting for the rapid proliferation and transmission of pathogens that would not have been able to survive so comfortably in the aforementioned scattered human societies. Thus, the growth and expansion of the human population during this period may have greatly favored infectious agents harmful to our species, giving rise to what was probably the first great wave of emerging human diseases. From then on, since only their harmful effects manifested themselves, our war against the microbial world began, and has persisted. Although we did not know what caused the epidemics that often depleted the population, some sources from the Ancient Ages already spoke of invisible germs that transmit diseases. During the Renaissance, Europe began to suspect that contagious diseases were caused by "living germs" that were passed in one form or another, from one individual to another. These early attempts to rule out the supernatural causes that had hitherto explained disease were reaffirmed by the arrival of syphilis in Europe, for which contact was clearly necessary for contagion.

It was not until the middle of the nineteenth century, after observing certain forms of parasitic life in animals and plants, that the direct implication of specific microorganisms in the appearance of diseases was demonstrated. By the end of the century, a large number of pathogenic infectious agents had already been characterized; at the same time, increasing progress was made in the knowledge of the mechanisms of transmission and asepsis. Finally, in the twentieth century, with the advent of antibiotics, vaccines and sanitary improvements, we were able to tackle infectious diseases. A revolution in the clinical world began, which allowed us to control and defeat the pathogens that endangered our survival: *antibiotic fever*. When scientists began to identify the microorganisms inside us, they also stumbled upon the ones that cause infectious diseases, so it was only logical that they all got a bad reputation. Although Metchnikoff described the benefits of consuming products fermented with lactic acid bacteria in the early twentieth century, they soon were declared the enemy, and a battle against these horrible disease-causing microbes began. Since then, we have managed to curb epidemics and reduce mortality associated with infectious diseases. However, should we fight them all without distinction? It is only now, after years of overuse of antibiotics, that we are beginning to realize the collateral damage of this war. When an infection is treated with an antibiotic, the antibiotic is not only directed against a specific bacterium but also destroys many other bacteria in addition to the pathogen. We are also eliminating the good guys, and wreaking havoc on our beneficial microbiota that have been with us since our origins, and on which we depend so much. Does this mean that they should be banished from current clinical practices? Far from it. Antibiotics are remarkable; they save lives at a price that sometimes it is undoubtedly necessary to pay, but it is essential to control their prescription and administration, resorting to them only when it is of vital importance.

Understanding how we are connected to the microbiota and our mutual interdependence is essential to neutralize infectious germs while protecting those that keep us healthy, avoiding as much as possible the harmful effects of broad-spectrum antibiotics that wipe out everything. In recent years, the possibility of treating diseases of different origins has been considered, not only without disturbing our "good" microbiota but also using its own beneficial potential to restore balance. This would require shifting the paradigm from killing bacteria to exactly the opposite: adding more bacteria. Thus far, many foods incorporate live microorganisms that produce a health benefit to the individual when ingested—*probiotics*. These are species that are found among the normal microbiota of the intestine, and have the ability to repair possible alterations in populations, to some extent balance the response of the immune system, or even stand up to pathogenic invaders. Future treatments for specific pathologies could be limited exclusively to the use of probiotics, but to be considered truly effective, further clinical trials are still needed to determine the strains, doses, and conditions of application.

Another future prospect in the treatment of various intestinal dysbioses is to replace the altered microbiota with one from a healthy individual. Since half of the weight of our fecal waste is made up of microorganisms, what better avenue to explore than to use a healthy, rich and diverse microbiome? Indeed, fecal matter transfer is a promising therapeutic alternative for many intestinal pathologies. Indeed, it is an option that is being considered to restore the alterations caused by the pathogen *Clostridium difficile in an environmentally friendly way* and has great therapeutic potential for other diseases, such as inflammatory bowel disease, autoimmune diseases, allergies, and obesity. The possible applications must be reconciled with good practices, but undoubtedly present a hopeful future perspective for the treatment of these and possibly other conditions.

We are beginning to realize that humans and microbes are one and the same entity in sync; for optimal health, we must pay attention to both our human and microbial sides. In keeping with other ecological demands, we adhere to the message "save our microbiota—the *good* one." Although not on a scale as large as the Earth, we also form an ecosystem that we must take care of as a whole. We are in contact with the internal and external microbial world. With industrialization and modern life, we have separated ourselves from nature and the balanced relationship that kept us in harmony. We have gone from a life where we were exposed to a myriad of different species, to living in aseptic environments with the belief that any invisible species is harmful and therefore worthy of extermination. The microbes that surround us have been forced to survive and adapt to an aggressive and artificial environment created by an abuse of disinfectants and antimicrobials, seriously affecting their composition and heterogeneity. It is not about living in dirty or unhealthy environments (that goes without saying), but an excess of cleanliness that alters our balance with beneficial microorganisms will provide a positive outcome. Asthma, inflammatory diseases, obesity, allergies, and other autoimmune disorders are diseases whose incidence has skyrocketed since the mid-twentieth century in developed societies. Increasing scientific evidence suggests that the lack of many bacterial species contributes to the increase in these disorders. Our immune system had evolved in such a way that it was trained to fight them, but without them, the mechanisms that modulate the immune response can become unbalanced, affecting the overall functioning of our defenses and other systems in different ways.

Let us look at them from now on with different eyes. Even though they are invisible, they are more important than we first thought, both externally and internally. Without them, life as we know it would be seriously compromised, and the conditions in which we live would change in ways we cannot even imagine.

# Bibliography

Alarcón Cavero T, D'Auria G, Delgado Palacio S, del Campo Moreno R, Ferrer Martínez M (2016) Procedures in clinical microbiology. Microbiota [Internet]. In: Cercenado Mansilla E, Cantón Moreno R (eds). Spanish Society of Infectious Diseases and Clinical Microbiology (SEIMC), Madrid [cited 2018 Nov]. Available from: https://www.seimc.org/contenidos/documentoscient ificos/procedimientosmicrobiologia/seimc-procedimientomicrobiologia59.pdf.

Annalisa N, Alessio T, Claudette TD, Erald V, Antonino DL, Nicola DD (2014) Gut microbiome population: an indicator truly sensitive to any change in age, diet, metabolic syndrome, and lifestyle. Mediators Inflamm 2014:901308

Atherton JC, Blaser MJ (2009) Coadaptation of Helicobacter pylori and humans: ancient history, modern implications. J Clin Invest 119(9):2475–2487

Bäckhed F, Ley RE, Sonnenburg JL, Peterson DA, Gordon JI (2005) Host-bacterial mutualism in the human intestine. Science 307(5717):1915–1920

Bokulich NA, Chung J, Battaglia T, Henderson N, Jay M, Li H, Lieber AD, Wu F, Perez-Perez GI, Chen Y, Schweizer W, Zheng X, Contreras M, Dominguez-Bello MG, Blaser MJ (2016) Antibiotics, birth mode, and diet shape microbiome maturation during early life. Sci Transl Med 8(343):343ra82

Briones C. The spark of life: Miller's experiment, sixty years later. [Internet]. [Retrieved Feb 2019]. Available from: https://naukas.com/2013/12/27/la-chispa-dela-vida-el-experimento-demiller-sesenta-anos-despues/

Clemente JC, Pehrsson EC, Blaser MJ, Sandhu K, Gao Z, Wang B, Magris M, Hidalgo G, Contreras M, Noya-Alarcón Ó, Lander O, McDonald J, Cox M, Walter J, Oh PL, Ruiz JF, Rodriguez S, Shen N, Song SJ, Metcalf J, Knight R, Dantas G, Dominguez-Bello MG (2015) The microbiome of uncontacted Amerindians. Sci Adv 1(3)

Coppini MV (2019) Habitat and ecological niche how do they differ? [Internet] [Retrieved Feb 2019]. Available from: https://geoinnova.org/blog-territorio/habitat-nicho-ecologico/

Costello EK, Stagaman K, Dethlefsen L, Bohannan BJ, Relman DA (2012) The application of ecological theory toward an understanding of the human microbiome. Science 336(6086):1255–1262

Davies J (2001) In a map for human life, count the microbes, too. Science 291(5512):2316

De Filippo C, Cavalieri D, Di Paola M, Ramazzotti M, Poullet JB, Massart S, Collini S, Pieraccini G, Lionetti P (2010) Impact of diet in shaping gut microbiota revealed by a comparative study in children from Europe and rural Africa. Proc Natl Acad Sci U S A 107(33):14691–14696

Dethlefsen L, McFall-Ngai M, Relman DA (2007) An ecological and evolutionary perspective on human-microbe mutualism and disease. Nature 449(7164):811–818

Dominguez-Bello MG, Blaser MJ (2008) Do you have a probiotic in your future? Microbes Infect 10(9):1072–1076

Dominguez-Bello MG, Peterson D, Noya-Alarcon O, Bevilacqua M, Rojas N, Rodríguez R, Pinto SA, Baallow R, Caballero-Arias H (2016) Ethics of exploring the microbiome of native peoples. Nat Microbiol 1(7):16097

Dominguez-Bello MG, Knight R, Gilbert JA, Blaser MJ (2018) Preserving microbial diversity. Science 362(6410):33–34

Gamiño-Arroyo AE, Barrios-Ceballos MP, Cárdenas de la Peña LP, Anaya-Velázquez F, Padilla-Vaca F (2005) Normal flora, probiotics and human health. Acta Universitaria 15(3):34–40

Genetic Science Learning Center. The Human Microbiome [Internet]. 15 Aug 2014 [Retrieved Nov 2018]. Available from https://learn.genetics.utah.edu/content/microbiome/

Ghose C, Perez-Perez GI, van Doorn LJ, Domínguez-Bello MG, Blaser MJ (2005) High frequency of gastric colonization with multiple Helicobacter pylori strains in Venezuelan subjects. J Clin Microbiol 43(6):2635–2641

Gilbert JA (2015) Our unique microbial identity. Genome Biol 16(1):97

Gilbert JA, Neufeld JD (2014) Life in a world without microbes. PLoS Biol 12(12):e1002020

Gilbert JA, Blaser MJ, Caporaso JG, Jansson JK, Lynch SV, Knight R (2018) Current understanding of the human microbiome. Nat Med 24(4):392–400

González VS, Delgadillo AA (2002) Cutaneous flora as protection and barrier of normal skin. Rev Cent Dermatol Parcua 11(1):18–21

Goodrich JK, Waters JL, Poole AC, Sutter JL, Koren O, Blekhman R, Beaumont M, Van Treuren W, Knight R, Bell JT, Spector TD, Clark AG, Ley RE (2014) Human genetics shape the gut microbiome. Cell 159(4):789–799

Gut Microbiota and Health Section of the European Society for Neurogastroenterology & Motility (ESNM) (2018) Diet & Gut Microbiota. The influence of diet on the microbiota [Internet]. [Retrieved Oct 2018]. Available from https://www.gutmicrobiotaforhealth.com/es/dieta-y-microbiota-intestinal/

Human Microbiome Project Consortium (2012) Structure, function and diversity of the healthy human microbiome. Nature 486(7402):207–214

Iáñez E (1998) General microbiology course. Concept and history of microbiology [Internet]. 17 Aug 1998 [retrieved Oct 2018]. Available from http://www.biologia.edu.ar/microgeneral/microianez/01_micro.htm

Koliada A, Syzenko G, Moseiko V, Budovska L, Puchkov K, Perederiy V, Gavalko Y, Dorofeyev A, Romanenko M, Tkach S, Sineok L, Lushchak O, Vaiserman A (2017) Association between body mass index and Firmicutes/Bacteroidetes ratio in an adult Ukrainian population. BMC Microbiol 17(1):120

Kumate J, Gutiérrez G, Muñoz O, Santos I, Solórzano F, Guadalupe M (2008) Clinical Infectology Kumate-Gutiérrez, 17ª edn. Méndez Editores, Mexico

Ley RE, Turnbaugh PJ, Klein S, Gordon JI (2006) Microbial ecology: human gut microbes associated with obesity. Nature 444(7122):1022–1023

Li K, Bihan M, Yooseph S, Methé BA (2012) Analyses of the microbial diversity across the human microbiome. PLoS ONE 7(6):1–18

Luisi PL (2010) Emerging life: from chemical origins to synthetic biology, 1st edn. Tusquets Editores S.A, Barcelona

Maldonado-Contreras A, Goldfarb KC, Godoy-Vitorino F, Karaoz U, Contreras M, Blaser MJ, Brodie EL, Dominguez-Bello MG (2011) Structure of the human gastric gastric bacterial community in relation to Helicobacter pylori status. ISME J 5(4):574–579

Mancebo A (2015) The human microbiota: how bad bacteria are! [Internet]. 18 Feb 2015 [retrieved Oct 2018]. Available from http://meetgenes.blogs.uv.es/la-microbiota-humana-que-malas-son-las-bacterias/

McFall-Ngai M, Hadfield MG, Bosch TC, Carey HV, Domazet-Lošo T, Douglas AE, Dubilier N, Eberl G, Fukami T, Gilbert SF, Hentschel U, King N, Kjelleberg S, Knoll AH, Kremer N, Mazmanian SK, Metcalf JL, Nealson K, Pierce NE, Rawls JF, Reid A, Ruby EG, Rumpho M, Sanders JG, Tautz D, Wernegreen JJ (2013) Animals in a bacterial world, a new imperative for the life sciences. Proc Natl Acad Sci U S A 110(9):3229–3236

Murray PR, Rosenthal KS, Pfaüer MA (2009) Medical microbiology, 6a edn. Elsevier-Mosby, Madrid

Pérez Mercader J (2018) Astrobiology Center associated with the NASA Astrobiology Institute. Astrobiology. "Life as a consequence of the evolution of the Universe" [Internet]. [Retrieved Oct 2018]. Available from http://www.cab.inta.es/es/astrobiologia

Relman DA (2012) Microbiology: learning about who we are. Nature 486(7402):194–195

Solomon EP, Berg L, Martin D (2013) Biology, 8th edn. McGraw Hill-Interamericana, Mexico City

Walter J, Ley R (2011) The human gut microbiome: ecology and recent evolutionary changes. Annu Rev Microbiol 65(1):411–429

Young VB (2012) The intestinal microbiota in health and disease. Curr Opin Gastroenterol 28(1):63–69

**Patricia Puerta Gil** has a double degree in Biology and Food Science and Technology from Universidad Autónoma de Madrid (UAM) and a master's degree in Microbiology for the Food Industry from Institut Universitari de Ciència i Tecnologia de Barcelona (IUCT). She currently works as a researcher at Instituto de Agroquímica y Tecnología de Alimentos (IATA-CSIC), in València. She obtained her Ph.D. in Food Science and Technology from the Universitat Politècnica de València (UPV)-IATA-CSIC, and her research work has been focused on sensory analysis and consumer science. Additionally, she has participated in several specialization programs in scientific dissemination strategies for researchers organized by Universitat Politècnica de València and IATA-CSIC.

# Bovine Belching as an Alternative Source of Energy

Alejandro Salgado Flores◉

When we think of alternative sources of energy production, the first things that come to mind are probably images of gigantic metal mills that harness the power of the wind to generate energy, photovoltaic panels that allow the conversion of solar energy into electrical energy, or obtaining energy directly from the geological activity of the Earth. However, there are other sources in nature of biological origin that also have great potential. To explain this in more detail, we need to change our everyday perspective of the world and adapt it to an infinitesimal scale—more specifically, to the world of the microscopic.

However, let us take it one step at a time. Let us first consider a ruminant, for example, a cow peacefully grazing while enjoying the warmth of the spring sunshine. Given its relaxed appearance, few would think of the frenetic biological activity occurring in its rumen (one of the four compartments into which the stomach of ruminants is divided) where the first and largest part of digestion takes place. Since it is a blind sac, the decomposition of the vegetable matter ingested and deposited in this compartment takes place practically in the absence of oxygen, that is, in anaerobic conditions. It is important to understand that plant cells contain a large amount of polysaccharides (i.e., polymers), or repeats of simpler carbohydrate molecules such as glucose. Depending on their composition and structural arrangement, there are various types of polysaccharides (cellulose, hemi-cellulose, pectin, starch), proteins, and lipids. These compounds are intertwined like a wicker basket, forming the structural part of the plant cell. Depending on the plant, the

A. Salgado Flores (✉)
Thermo Fischer Scientific, Madrid, Spain
e-mail: alejandrosalgadoflores@gmail.com

External Collaborator, Department of Arctic and Marine Biology, UiT-The Arctic University of Norway, Tromsø, Norway

type of polysaccharide and its proportion will vary, which will affect its digestion as we will see.

Due to the nature and intertwined structure of these polymers, digestion (i.e., the breakdown or degradation of these compounds into their simpler constituents) would be an impossible task for ruminants if they had to do it on their own. Similar to other mammals, ruminants lack the capacity to produce the enzymes necessary for digestion, from which the energy and nutrients necessary for the animal's metabolic activities (respiration, circulation, tissue regeneration, etc.) are obtained. Therefore, if ruminants cannot digest these compounds by themselves, how is it possible that cows can live exclusively on the ingestion of kilos and kilos of a food as indigestible as grassy vegetation? This is due to symbiotic associations between the host and a very diverse community of microorganisms—called microbiota—specialized in the decomposition of the polysaccharides present in their diet. It is precisely the activity of this microbiota that allows ruminants to act as true biological harvesters. As the saying goes, there is strength in numbers; in this case, those numbers refers to the cow's diverse microbiota, which enables complicated digestion to take place.

As observed in other natural environments such as soil, oceans, plant roots, or skin, the microbiota present in the rumen is made up of different groups of microorganisms, interacting with each other in a harmonious way to carry out the degradation of complex compounds into simpler ones. This microbial Tower of Babel is composed of multiple groups with different culinary tastes, each specialized in the degradation of different polysaccharides from which the energy to meet their needs is obtained. Groups of bacteria (prokaryotes),[1] protozoa and fungi (both eukaryotes) are specialized in the degradation of the most complex structural parts by using protein scissors that allow them to cut polysaccharides into their smallest compounds. These compounds are, in turn, metabolized through multiple cycles of decomposition by other microbial groups (bacteria and protozoa) to their simplest elements. This microscopic recycling chain allows structural compounds as indigestible as cellulose to degrade into their most elementary particle: glucose molecules.

From the point of view of the microorganism, the decomposition of these polymers allows it to obtain the energy and nutrients necessary for its biological activities. However, what benefits do ruminants derive from the digestion of plant matter by the microbiota? It is not all about carbohydrates. Specifically, as a result of the anaerobic degradation of these plant polymers, a series of compounds called volatile fatty acids (VFAs) are produced; VFAs are biological molecules with lipid structures soluble in fats or oils (liposoluble). Since the walls of the rumen are constituted of animal cells whose surfaces are made up of lipid structures, VFAs are easily absorbed, accessing the circulatory system and finally being

---

[1] Prokaryotes and Eukaryotes constitute (along with Archaea) the main taxonomic classification domains in biology. Currently, biological organisms are classified on the basis of their genetic similarity using universal genetic markers.

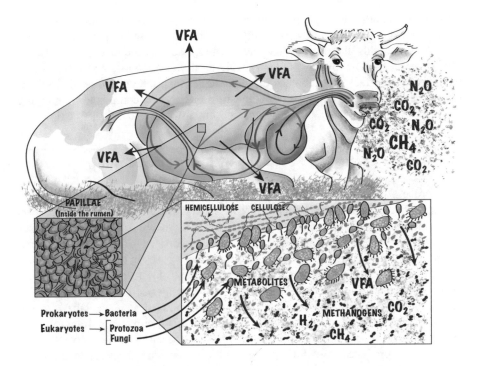

**Fig. 1  Anaerobic decomposition of plant matter by the microbiota present in the rumen**. The foregut of ruminants consists of four chambers, with the reticulum (red) and rumen (green) merged into a single organ. Plant particles are physically broken down in multiple chewing cycles (direction of arrows) to facilitate their breakdown by the microbiota (Boxes 1 and 2). Fungi and some species of protozoa and bacteria are attached to plant particles, and are responsible for the physical breakdown of polysaccharides (cellulose, hemicellulose). The resulting compounds are, in turn, degraded into simpler elements by other species of bacteria and protozoa floating in the rumen liquid (lumen). Some products resulting from this microbial decomposition, such as volatile fatty acids (VFAs), are absorbed by the rumen wall and incorporated into the animal's metabolism, and used as a source of energy. The last steps of anaerobic decomposition are mediated by methanogens (archaea), which produce mainly methane ($CH_4$) as a waste product. This methane, together with other gases, such as carbon dioxide ($CO_2$) and nitrogen oxide ($N_2O$), is expelled to the outside by the animal, mainly through the mouth

used by the host as energy sources for its metabolism. In ruminants, up to 70% of energy requirements are covered by VFA assimilation (Fig. 1).

There are several types of VFAs produced by the microbiota that contribute in different ways to the metabolism of the host. One example is propionic acid, which is the main energy substrate in ruminants for the production of glucose without using carbohydrates; in humans, butyric acid is absorbed by colon cells (colonocytes), and has been recognized to have protective properties against colon cancer, for example. The microbiota also provides the host with amino acids which build proteins, or protect against the harmful effects produced by some plant compounds, such as the secondary metabolites of plants.[2] These are perfect examples of how two worlds, the microscopic and the macroscopic, interact to mutual benefit: one (the host) providing food as well as a warm, moist, and oxygen-free environment, perfect for decomposing microorganisms; the other (the microbiota) doing the dirty work by allowing the digestion of complex compounds, supplying the host with proteins and energy sources essential for its biological functions.

## The (Large) Ecological Footprint of Livestock Farming

This process of microbial degradation, however, is not perfect: there are waste products that cannot be reused and must be excreted to avoid their accumulation inside the rumen, which would be detrimental to the health of the animal. Gases (such as carbon dioxide, methane, nitrogen, or hydrogen sulfide) are one such example. In extreme cases, the accumulation of these gases in the rumen can eventually cause the death of the animal by asphyxiation due to excessive distention of the abdominal wall, which puts pressure on the lungs. Of these gases, methane makes up approximately 25% of the total, and is produced by a specialized group of microorganisms called methanogens.[3] Although this group is important to maintain a correct process of anaerobic degradation in the animal's digestive system, the methane produced is of little energy value to the animal and/or to the other groups of microorganisms present in the rumen, making it necessary to expel it to avoid its accumulation. This means that ruminants, and especially cows used for milk production, are real factories of methane, expelling up to 300 L a day into the atmosphere![4] Interestingly, approximately 85% of the methane produced in the rumen is directly expelled through the animal's mouth, and the remaining 15% is expelled through the rectum. Therefore, contrary to common belief, it is not cow "wind" that is the major contributor to the accumulation of atmospheric methane but, in fact, bovine belching. Regardless of the expulsion mechanism of these gases, methane produced directly by livestock activity[5] constitute approximately 18% of total global methane emissions, and represent 10% of total greenhouse gas (GHG) emissions from human activities.[6]

---

[2] These are chemical compounds produced by plants without a direct vital function. In many cases, these compounds may have an antimicrobial function, or because of their unpleasant taste they have a deterrent effect on herbivores that feed on the plant in question.

[3] Methanogens are taxonomically classified as archaea.

[4] The amount of methane produced per animal can vary depending on the volume of the rumen, the type of diet, and the genetic makeup of the animal.

[5] Methane emissions from industrial activities involving cattle and goats (ruminants), as well as poultry, have been reported. However, the bulk of methane emissions come from cattle.

[6] The remaining GHGs from human activities correspond mostly to carbon dioxide (81%), mainly from transport, industry, and deforestation. Other GHGs of global importance are nitrogen oxides (6%) and fluorinated gases (3%). Source: https://www.epa.gov/global-mitigation-nonco2-greenhouse-gases/global-anthropogenic-nonco2-greenhouse-gas-emissions.

Of all these gases, carbon dioxide holds the title of global emissions champion, accounting for approximately one-third of the total; however, methane gas is becoming increasingly important in assessing the environmental impact capacity of the accumulation of GHGs in the atmosphere. This is because its molecular structure allows methane to trap heat 25 times more efficiently than, for example, the same amount of carbon dioxide molecules. Considering this, although methane is emitted to a much lesser extent than carbon dioxide, the potential effect of its atmospheric accumulation is currently considered key to better understanding global warming.

In recent decades, large resources have been invested in developing strategies to mitigate methane emissions produced directly by livestock activities. Plant-based supplements have been included in their diets, such as *phenolic compounds*, a group of the aforementioned secondary plant metabolites, which significantly reduce the growth of methanogens and other microorganisms directly associated with methane production in the rumen. Additionally, experimental vaccines have been developed to train the ruminant's immune system to selectively eliminate specific groups of methanogens. However, neither strategy is free of complications. Phenolic compounds can be highly toxic in large concentrations; severe liver damage or even death has been described in some cases where animals have ingested plants with a high content of these compounds. In the case of vaccines, results obtained are not very consistent. The variability and diversity of methanogens, due to factors such as the diet and genetics of the animal, limit the effectiveness of vaccines.

For all these reasons, currently, the main strategy to mitigate livestock-derived methane emissions is the manipulation of dietary intake. This is related to the fact that the type and proportion of polysaccharides ingested can have a substantial impact on the composition of the rumen microbiota, i.e., which groups of microorganisms are present, and in what proportion. For example, a higher intake of structural fiber, such as cellulose or hemicellulose, favors methane production by promoting the creation of microorganisms that decompose the compounds needed by methanogens to produce methane (hydrogen, carbon dioxide, acetic acid and methylamine compounds). Whereas, the ingestion of plants rich in soluble polysaccharides (e.g., starch or pectin) require shorter digestion time in the rumen, limiting the production of methane; other compounds used by microbial groups competing with methanogens (and which do not produce methane) are produced instead. In short, we humans have not only become obsessed with controlling our diet, but we devise dietary plans for our animals. Is this a crazy world, or what?

Considering this, it would seem that simply feeding cattle a diet composed exclusively of material rich in soluble polysaccharides would be enough to reduce methane emissions; however, it is not that simple. Consuming only feed that is rich in soluble compounds can be very detrimental to the animal's health by making the rumen fluid more acidic, and increasing its density; that, in turn would make it difficult to expel the gases resulting from microbial digestion. In extreme cases, this could lead to asphyxia and death.

## Methane as a Waste Substance?

As we have just seen, strategies to mitigate methane production have had mixed results, and are not without risks to animal health. However, is it absolutely necessary to achieve zero emissions of methane produced by ruminants? Would it not be better to utilize the high energy capacity of the methane molecule as an energy source? In this way, a reduction of methane emissions into the atmosphere could be achieved, while also harvesting an alternative energy source.

The saying "if you can't beat 'em, join 'em" has been the mantra of numerous research groups over the last decades. A clear example is the biogas plant: waste products such as animal feces, food scraps, or plants are stored in large silos at high temperatures (above 38 °C) to facilitate anaerobic digestion by microorganisms similar to those present in the rumen, eventually producing methane along with other gases such as hydrogen sulfide. This methane is stored for subsequent combustion, producing electricity and heat.

Following this approach, there have been individual initiatives to adapt the concept of biogas plants to more everyday uses. Here is an ingenious example: the English inventor Brian Harper has designed small-scale anaerobic digesters intended to be installed in lampposts, and used as containers to dispose of dog waste. The resulting gases (methane and hydrogen sulfide) from the decomposition of the fecal matter is filtered, and accumulates in the upper part of the container and powers the lights during the night hours. By applying a little ingenuity, what a priori was a waste management problem has been transformed into an innovative way of generating electricity for use in urban lighting. As the famous *Law of Conservation of Energy* states, "energy can neither be created nor destroyed—only converted from one form of energy to another." Whether the scientists who helped define that universal law were thinking of the energy potential of feces is open to speculation.

However, although efficient for animal waste management, these strategies still do not solve the problem of methane emission directly by cows when digesting their feed, as most of the methane is expelled directly from the animal's mouth. The question: is there a way to make use of all the methane produced by cows that is otherwise wasted and emitted into the atmosphere? With this question in mind, researchers at the National Institute of Agricultural Technology (INTA) in Argentina have developed a simple but inspiring solution for using the methane produced in a cow's rumen. The idea: by using a plastic cannula (tube) inserted directly into the upper part of the rumen, where a greater amount of gases are stored, these gases can be channeled into a plastic backpack attached to the animal. As the animal digests the food, the gases that are produced, including methane, are stored in this backpack; at the end of the day these gases can be collected for subsequent purification. In the purification process, the different gases, such as carbon dioxide and hydrogen sulfide, are separated to finally obtain the coveted biomethane, which is then compressed for easy transport and handling. However, how much energy could be generated from the methane produced by a single cow in a day? Considering a methane production of approximately 300 L per day, the energy contained in this volume of gas could be used for many different purposes. For example, it has been demonstrated that with the amount of methane obtained from a cow, it is possible to run a refrigerator for a whole day; the Argentine scientists from INTA, were even able start and drive a car with some autonomy. Although still far from large-scale application, perhaps it is not unthinkable that in the future we could see meadows inhabited by backpacker cows acting as biological factories of methane gas as an alternative source of energy.

## Methane Eating Bacteria

As in every great story, the main character always has a nemesis, an antagonistic character. In the world of microorganisms, and in the production of methane specifically, this is no less true. On the one hand, we have methanogenic bacteria; on the other hand, there are also bacteria capable of metabolizing methane and using it for their own benefit. Let me introduce you to the group of methane-eating bacteria: *methanotrophs*. This group of bacteria generally grows in the presence of oxygen, but can also occur in environments devoid of it. What is certain is that this group of microorganisms is present in the vast majority of ecosystems where methane is produced in a significant way: from the ocean floor to rice fields, marshes, peat bogs, and lake environments. From a global point of view, methanotrophic bacteria constitute the main sink of atmospheric methane of biological origin. In other words, the activity of this group of bacteria controls the volume of atmospheric methane, thus preventing its uncontrolled accumulation, which would further impact the planet's climate. How do these microorganisms eat methane? In the presence of oxygen, methanotrophs are able to exploit methane gas thanks to the specific activity of an enzyme called methane monooxygenase, which allows the addition of an oxygen molecule to the methane molecule, oxidizing it. Due to this oxidation process, methane is transformed into a different compound, which is finally assimilated by methanotrophs for metabolism. We can deduce, therefore, that an inert compound such as methane constitutes a perfect food for these bacteria to nourish and proliferate.

On the other hand, in oxygen-poor environments, such as seabeds, peat bogs or marshes, methane molecules cannot be oxidized in the aforementioned way. However, nature is wise and has given methanotrophs[7] in these environments the ability to metabolize methane by using other compounds derived from metals such as manganese, or elements such as sulfur (sulfates) or nitrogen (nitrates). For example, only in marine environments has the oxidation of methane by this route been estimated to constitute approximately 80% of the methane produced and, therefore, avoid a large accumulation in the atmosphere. However, what about in the rumen? It is a closed environment where a large amount of methane is produced, so it has

---

[7] Although in an aerobic environment methanotrophs are classified within the *Bacteria* domain, in environments devoid of oxygen, i.e., anoxic or anaerobic, methane-oxidizing bacteria belong to the *Archaea* domain, as do methanogenic bacteria.

everything necessary for methanotrophs to grow happily. However, the reality is quite different: methane oxidation in the rumen actually occurs in a very limited way. This impediment is mainly related to the exquisite metabolism of methanotrophs, as they depend on very specific and complex nutritional requirements for their growth. Therefore, participation of other microbial groups present in a low proportion or absent in the rumen is essential. In addition, these other microbial groups are better adapted to the environment of the rumen, and further limits the growth of methanotrophic bacteria and methane oxidation.

## How to Convert Methane into Food and Energy

The great limitations in the growth of methanotrophs in the rumen make their use as a strategy to reduce livestock-produced methane not very plausible. However, perhaps the peculiar metabolism of these microorganisms can be exploited for other industrial purposes. For example, in recent decades, the ability to produce food from methane oxidation has been investigated. Methanotrophs have the ability to produce organic waste products from methane oxidation, which could be used to produce protein-rich foods. The potential is such that some companies have begun mass production to manufacture animal feed; there is even speculation that it could be used to produce food for human consumption. In the event that this solution is finally productive, it is intriguing to consider that methane produced by the digestion of livestock feed could, in turn, be used to produce food for these same livestock, thus closing the ecological cycle and reducing its environmental impact.

However, the main industrial application of methane conversion by methanotrophic bacteria is focused on the sustainable production of biofuels, such as bioethanol. Remember, these bacteria oxidize methane to produce organic compounds, which are subsequently used in their own metabolism.[8] By treating these with chemical compounds, the desired biofuel can be obtained. In addition to being a useful strategy to mitigate the accumulation of atmospheric methane, the production of biofuels through the activity of methanotrophic bacteria can reduce the impact of traditional production models for obtaining this type of fuel. Currently, the production of biofuels, such as bioethanol, is based on the fermentation (i.e., the conversion of one organic compound into another without the presence of oxygen) of plant sugars present in crops such as sugar beet or sugar cane. Despite having a lower environmental impact after combustion compared to gasoline or diesel, the large-scale production of this type of biofuel requires using large amount of arable land to grow the necessary amount of sugar beet, corn,

---

[8] From the point of view of thermodynamics, a calorie (Cal) is the energy required to raise the temperature of one gram of water by one scientific degree. The metabolism of one gram of protein generates four kilocalories (kCal) or four thousand calories; one gram of lipids generates 9 kcal. This is why diets rich in lipids are more fattening, as they contain more energy, and hence lipid compounds will be better as biofuels.

or palm. This, however, causes collateral damage when lands such as forests and jungles are used, accelerating deforestation. In some countries, such as Brazil and Indonesia, deforestation due to the production of vegetables for biofuel production is a major environmental problem. If we are able to bring this strategy to large-scale production in a more efficient way, perhaps, one day, the role of this group of microorganisms unknown to the general public can play a key role in global energy strategy. It will be interesting to see what the future holds.

## Bibliography

Dehority BA (2003) Rumen microbiology. Nottingham University Press, Nottingham

Fei Q, Guarnieri MT, Tao L, Laurens LM, Dowe N, Pienkos PT (2014) Bioconversion of natural gas to liquid fuel: opportunities and challenges. Biotechnol Adv 32(3):596–614

Fleming N (2018) From stools to fuels: the street lamp that runs on dog do. The Guardian (International Ed.). 1 Jan 2018; Sect. News: Environment (Energy). Available at: https://www.the guardian.com/environment/2018/jan/01/stools-to-fuels-street-lamp-runs-on-dog-poo-bioene rgy-waste

Global Anthropogenic Non-$CO_2$ Greenhouse Gas Emissions: 1990–2030 [Internet]. United States Environmental Protection Agency (EPA); 2012 [updated 9 Aug 2016]. EPA 430-R-12-006. Available from: https://www.epa.gov/global-mitigation-nonco2-greenhouse-gases/global-anthropogenic-nonco2-greenhouse-gas-emissions

Hanson RS, Hanson TE (1996) Methanotrophic bacteria. Microbiol Rev 60:439–471

Hook SE, Wright AD, McBride BW (2010) Methanogens: methane producers of the rumen and mitigation strategies. Archaea 2010:945785

Le Page M (2016) Food made from natural gas will soon feed farm animals—and us. New Scientist. 16 Nov 2016; Sect News: 3100 (issue). Available at: https://www.newscientist.com/article/2112298-food-made-from-natural-gas-will-soon-feed-farm-animals-and-us/

Reeburgh WS (2007) Organic methane biochemistry. Chem Rev 107:486–513

Scientific News and Information Service (sync). The flatulence of cows can power a car engine. ABC NEWSPAPER (National Ed.). 16 Oct 2013; Science Section. Available at: https://www.abc.es/ciencia/20131016/abci-flatulencias-vacas-pueden-alimentar-201310161253.html

**Alejandro Salgado Flores** holds a degree in Biology from Universidad Autónoma de Madrid (UAM), a master's degree in Microbiology and Parasitology-Research and Development from Universidad Complutense de Madrid (UCM) and a Ph.D. in Natural Sciences from the Arctic University of Norway (UiT), where he also worked as a teaching assistant in the Department of Marine and Arctic Biology. He has made several stays in research centers both nationally and internationally, and his research work has been focused on the study of the intestinal microbiota in animal models of Arctic latitudes to determine the influence of diet on the different microbial groups in relation to methane emissions. After working at *Aratech Lifestyle Technology*, a digital consulting company where he was responsible for the Research and Development Department, he is currently working at *Thermo Fischer Scientific*, a biotechnology company where he works as technical application specialist for sequencing systems.

# The Secret Life of Mushrooms

Antonio González Ruiz◉

More than two million species have been described on planet Earth, from microscopic beings to larger organisms, such as humans. Obviously, with such a large number of organisms, the differences are not subtle. That is why throughout history multiple ways have been developed to categorize and define the organisms that live on the planet. One of the simplest, and most basic, ways to classify all organisms is to do it in five kingdoms. It is a technique that groups organisms according to their most general characteristics. In the first kingdom, we find the simplest organisms, formed by a single cell, the bacteria; they are the moneras kingdom. In the second are protozoa and algae, also known as protoctists. The third corresponds to the fungi kingdom, whose characteristics we will discuss in this chapter. The fourth is the plant kingdom and, as its name indicates, corresponds to plants. Finally, the fifth and largest kingdom is the one into which we humans fall: the animal kingdom.

As we have said, this technique groups all living things into five groups according to their general characteristics, so, for example, if we look carefully in the Animal Kingdom, we will see that arachnids, fish, amphibians, reptiles, birds and mammals are grouped there, without taking into account obvious morphological and physiological differences. The differences and shortcomings in the organization of the kingdom system are also evident at the microscopic level: in the first

A. González Ruiz (✉)
Area of Pharmacology, Nutrition and Bromatology, Department of Basic Health Sciences, Unidad Asociada de I+D+i al Instituto de Química Médica (IQM-CSIC), Universidad Rey Juan Carlos, Madrid, Spain
e-mail: antonio.gonzalezr@urjc.es

High Performance Experimental Pharmacology Research Group (PHARMAKOM), Universidad Rey Juan Carlos, Alcorcón, Spain

kingdom, we find bacteria, organisms formed by a single cell and therefore simpler at the morphological level; however, there are also unicellular fungi classified in the third kingdom. What makes them different? If they have only one cell, why are unicellular fungi not also located in the first kingdom? The answer is that the levels of cellular organization and the capacity for environmental adaptation of bacteria and fungi are completely different. This difference can be seen very well in the genetic material (where all the information for the correct development of the cell is encoded) found floating in the cell cytoplasm, totally unprotected. In addition, the structures responsible for more specific functions (e.g., obtaining energy, or synthesizing proteins for the repair of cellular structures degraded by the passage of time) are much smaller and simpler than those of other more developed cells, leading them to be scattered throughout the cell cytoplasm.

All bacteria are included in a larger group: prokaryotic cells. In contrast, the rest of the cells are called eukaryotes; they have their genetic material located and protected in a nucleus (a compartment surrounded by a membrane), and their organelles are larger. Additionally, eukaryotic cells associate with each other, taking advantage of the fact that they can specialize and perform specific functions to form an organism. Human beings are made up of different types of cells. Some of these cells transmit electrical impulses, which carry information from the environment to the brain, or carry oxygen from the lungs to the rest of the body. Clearly, there are many differences between organisms, but we all have one thing in common: we need energy to live.

## Energy Is Neither Created Nor Destroyed

Energy is basic to life; we need it to move, breathe, think, pump blood, digest, regulate body temperature, and vital functions in general. Each organism has devised a different strategy to obtain that energy. There are bacteria that live in volcanoes at the bottom of the Atlantic Ocean because they have developed mechanisms that obtain energy from the chemical compounds released by those volcanoes. Plants use the energy that comes from the sun to make nutrients (carbohydrates, fats, and proteins) and grow. According to the first principle of thermodynamics, energy can be stored and exchanged but can neither be created from nothing nor destroyed. That is why living things have found ways to store energy so that it can be accessed easily, quickly and safely.

One way of storing energy is in the chemical bonds that join atoms together to form molecules; when the molecules are broken, the energy contained in the bonds is released. However, not just any molecule will do because, depending on the type of molecule, more or less energy must be invested to break it. The stability of the bond between atoms dictates the amount of energy required: the more stable the bond, the more energy must be used. Therefore, it is logical to think that there are molecules that are broken using very little energy, but *when* broken give off enough energy to propel a car. For example, gasoline is composed of molecules that begin to break with the application of a small amount of external energy, giving rise to reactions that release a large amount of energy; that is why we use it as *food* for cars. In this respect, living beings are like a car designed by Mother Nature, which means that we need our own *gasoline*: nutrients.

There are two main types of nutrients: macronutrients, which are further divided into three subgroups (carbohydrates, lipids and proteins), and micronutrients (vitamins and minerals). Carbohydrates and lipids have the main function of providing energy to the body, while proteins mainly provide amino acids, molecules that replenish the components of the body's aging structures. They are also part of the millions of chemical reactions that constantly occur in our body. Vitamins and minerals do not provide energy, but are essential for all these reactions to function properly.

All organisms have evolved mechanisms to ensure they always have enough nutrients to survive. Autotrophic organisms such as plants, for example, produce their own nutrients. With their leaves, they capture the energy coming from the sun's radiation, and use it mainly for the development of their vital functions. However, part of that energy is stored in the form of carbohydrates and lipids for those times when there is no sun—remember, there are parts of the planet that have very few hours of sunlight, or none at all, for months. Other living beings, such as herbivorous, carnivorous, and omnivorous animals, digest complex molecules and obtain energy by breaking the bonds of the nutrients they ingest. In essence, they are factories that create very complex structures to carry out specific tasks. For example, tree trunks are made up of long, complex molecules that help them achieve greater height and deeper roots. The former makes them good competitors for sufficient sunlight, and the latter for water. The construction of these large and stable molecules stores energy, keeping these compounds (nutrients) unchanged for centuries, until they separate again into simpler molecules.

In another example of natural adaptation, one group of organisms found its source of nutrients in other dead organisms, animal skeletons, feces, branches, trunks, and leaves. This group is in the third kingdom, and is known as fungi. Without the action of fungi, the nutrients found in these materials would not be available for use by new generations of organisms, and life would eventually cease. Although it may not seem like it, the fungal method of feeding is very important to life on Earth because without this continual decomposition, essential nutrients would remain locked up in huge molecules and would not be available for use again by other organisms. This is why fungi are often referred to as nature's *funerary organisms*, as they take care of the tasks of decomposition and putrefaction of organisms that have died.

To degrade complex molecules, fungi secrete enzymes that reduce the amount of energy that must be used to separate the atoms that form the molecules. This way they obtain nutrients to continue growing. Using this strategy has worked very well for them. In fact, the record for the world's largest organism belongs to a specimen of the fungus *Armillaria ostoyae,* also known as honey-colored Armillaria. It lives in Oregon and spreads underground over an area of 9 km$^2$. Although approximately 100,000 species of fungi have been discovered to date, mycologists estimate that approximately one trillion more species have yet to be discovered.

With the technological resources we have today, it may seem incomprehensible that such a large number of species have not been discovered, but fungi are very versatile organisms, obtaining energy in the most unexpected places and in very different ways. For example, they have been allied for millions of years with members of other kingdoms, such as bacteria, algae, and plants, in a relationship that is very beneficial to both species. These associations or symbioses are quite common alliances in nature, as in the case of sharks and remora fish.

Associations of fungi with algae and bacteria are known as lichens. The fungus goes inside the algae to obtain sugar, and in exchange, they provide minerals sequestered in mineral structures and rocks. For this reason, lichens are among the first living beings to colonize newly formed volcanic islands and inhospitable places such as the Arctic, deserts and bare rock, although it must also be said that in these environments, their growth is very slow: Arctic lichen colonies spread at a rate of 2.5 to 5 cm every 1000 years. Despite their slow growth, lichens are extraordinarily long-lived; some are more than 4000 years old.

*Mycorrhiza* is another type of very successful association between fungi and plants. More than 5000 species of fungi are known to grow in close association with 80% of all rooted plants, including most trees. The success of this association has been so important for fungi and plants that it has been maintained for 460 million years. In fact, fossil records show that fungi and plants invaded the land from the sea in the same time period; these fossils show root structures similar to those that form today in the presence of mycorrhizae. The association consists of the fungus surrounding the root of the plant and inserting parts of its body into them, so the fungi receive energy-rich sugar molecules produced by the plants through photosynthesis, and which pass through their roots to the fungus. In return, the fungus breaks down organic nutrients so that the plant can use them to make sugars, and absorb minerals from the soil; some minerals are passed directly into the root cells. Phosphorus and nitrogen, key nutrients for plant growth, have been shown to be among the molecules that mycorrhizae carry from the soil to the roots. The fungi also absorb water and transfer it to the plant, which is an advantage in dry soils because it increases the plant's resistance to drought.

However, fungi are also the cause of most plant diseases. This is the case for American elm and American chestnut. Years ago, they were part of the landscape of the United States but, due to fungal infections that caused Dutch elm disease and chestnut blight, they were destroyed on a massive scale. From an economic and humanitarian perspective, fungi can cause enormous losses; they have been shown to have devastating effects on food supply throughout the world. Plant pests are particularly destructive, and cause billions of dollars of damage each year to cereal crops such as rice and wheat, which are staple foods for millions of people around the world. In addition, their characteristics and adaptability allow them to survive in stored grains, construction materials, and natural fibers, causing incalculable damage. This is the case in the timber and textile industries (e.g., cotton and wool), and is especially problematic in hot and humid countries such as India, as these conditions are conducive to fungal growth.

In summary, the survival of living beings depends on synthesis: cells or organisms break down the nutrients found in food, and store the energy that is released. That energy is stored again but in the form of a molecule called adenosine triphosphate (ATP), which is very easily broken down, allowing the cell to obtain energy quickly and easily. This work of destruction and construction is fundamental because the only way to access the cells' stored energy is through the breaking of the bonds that form ATP.

Since ATP is essential for the survival of the cell, ATP production was optimized. Cells developed a series of chemical reactions within their mitochondria to produce ATP in a chain. The chemical reactions are perfectly tuned, and produce up to 36 molecules of ATP from one molecule of glucose, a simple carbohydrate. Oxygen ($O_2$) is a crucial part of the reaction that achieves this amount of ATP. It accepts the electrons that move in the chain reaction, and by taking those electrons into its atomic structure, alters its atomic structure and binds to two carbon molecules, forming a new molecule: carbon dioxide ($CO_2$). However, in nature, there are situations where there is no $O_2$ in the environment, so many organisms living in places where this gas is not present, or may disappear, have developed what could be called emergency routes, so that in case there is no oxygen in the environment, they can continue to produce ATP. This type of cellular respiration in which oxygen is not present is known by the technical term *anaerobic respiration*, but in industry, it is called fermentation (Fig. 1). The term fermentation was first defined by Louis Pasteur as life in media devoid of oxygen. Fermentation, therefore, is a process for obtaining ATP when there is no oxygen present in the medium. The effects of not having oxygen to obtain ATP are that, on the one hand, only two ATP molecules are obtained; and, on the other hand, the resulting waste products change. In aerobic respiration, $CO_2$ and water are produced from the carbohydrates contained in food, while in alcoholic fermentation $CO_2$ and ethyl alcohol (both the alcohol that we drink and that we use to disinfect wounds) are produced.

Fermentation has been and is a fundamental process in the history of human development. It may seem contradictory to talk about fungi and increasing the shelf life of food, since fungi are able to degrade organic molecules very quickly, but the changes introduced by these microorganisms have been and are fundamental to extending the shelf life of food, and making it more digestible. Today, fermented foods of both animal and plant origin are an essential part of the diet of people in many countries. This is because microbial growth in food, both from naturally occurring and human-introduced populations, produces changes in those foods. These processes have been the main ingredient in the preparation of fermented foods in all cultures and places around the world, such as sauerkraut in northern Europe or soy sauce in the Far East. They have been so important that in cultures such as Indonesian, we can find a specific word for the inoculum: *ragi*, which means "starter," and refers to the inoculum used to start different types of fermentations. For example, Indonesian roti bread is made with a preparation of baker's yeast called *ragirotti*. In many cases the consumption of these foods is concentrated in the area where it historically has been consumed, in other cases

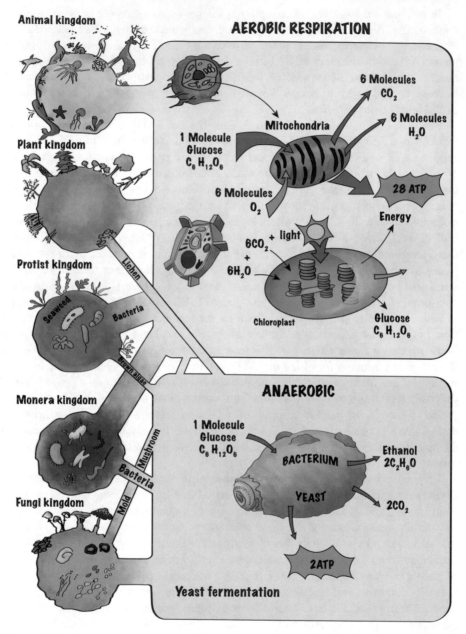

**Fig. 1 Differences between aerobic and anaerobic respiration.** On the left side of the image are the five kingdoms. Of these, only the animal and plant kingdoms can obtain energy through aerobic respiration, which is associated with the use of oxygen. The chloroplasts present in plants are responsible for the production of oxygen and glucose. While the kingdoms of protists, moneras, and fungi, have both aerobic and anaerobic routes, also known as fermentation, in this case, alcoholic anaerobic respiration is represented, in which ethanol is produced. In the plant kingdom, it is observed that lichens, an association between plants and fungi, are also able to carry out anaerobic respiration

its consumption has been globalized and are foods that we find in the supermarket like bread, beer, wine, cocoa, etc.

## Yeasts as a Source of Food Production and Transformation

Yeasts are unicellular fungi that began to be used because in prehistoric times, food preservation was ignored, and food was contaminated with microorganisms, particularly in the making of alcoholic beverages and bread. Later, through observation, humans were able to reproduce yeasts on demand. Their role in alcoholic fermentation was not known until the work of Pasteur in 1866–1876. It took some time to determine how these organisms survived without oxygen, but after that, their ease of cultivation, and the harmlessness of a large number of species have made them the most industrially used microorganisms for the production of alcoholic beverages, baked products, industrial alcohol, and glycerol. However, although yeasts are interesting organisms exploited for their own potential or as a support for new properties, they continue to be agents of undesirable alteration of food products if their development is not controlled. Think of moldy bread or rotten fruit.

### Bread Making

The best known and most widely used yeast is *Saccharomyces cerevisiae,* also known as baker's yeast. This microorganism is used to make common everyday foods such as bread, beer, and wine. Bread is one of the oldest traditional fermented foods and, in various forms, has been a fundamental constituent of the diets of many population groups for centuries. The history of bread dates back some six millennia to ancient Egypt, and depictions from this time illustrate in great detail the use of yeast to "rise" bread. Archaeological remains of bakeries have been found in many settlements of the time, including one of the most interesting for its location: at the base of the pyramid of Giza. Its remains date back to 2575 BCE, and it is estimated that it supplied bread daily to approximately 30,000 people. Other excavations have found samples of bread dating back to 2100 BCE.

There are few ingredients used in bread making; however, each one fulfils a fundamental function. The main ingredients are water, flour, salt and yeast. During fermentation,. the enzymes present in the yeast cause the carbohydrates contained in the flour to degrade into smaller molecules (sugars). This, in turn, allows the fungus (the yeast) to use them, producing $CO_2$ and other compounds derived from a fermentation process, which contribute to the bread's ultimate taste and nutritional value. Bakers can produce special breads using more complex mixtures of microorganisms. For example, the yeast *Saccharomyces exiguus* together with a bacterium of the genus *Lactobacillus* produces a characteristic sour taste and aroma.

## Beer Brewing

While it is true that bread was the food that could be found on the table in Egyptian homes, it is no less true that the drink that accompanied it was often beer. The beginning of beer brewing also originated at approximately 6000 B.C. in the Sumerian civilization and, according to records, its origin is parallel to the appearance of the first known agricultural implements. Most likely, humans discovered how to make beer by mistake. The most widely accepted hypothesis is that barley was left to soak for another purpose, and pieces of bread dough fell into the soaking barley. When this preparation was covered and in the absence of oxygen, a fermentation process began, eventually resulting in a barley wine. Beer is produced when yeasts such as the well-known *Saccharomyces cerevisiae* ferment simple carbohydrates from cereal grains (barley, wheat, corn, rice, etc.). The oldest formula for making this drink was found in southern Mesopotamia, and is preserved in the Metropolitan Museum in New York. Moreover, evidence has been found that in Mesopotamian culture, there was a cult of a brewing deity, which later evolved in Greek culture as the cult of the god Dionysus, and in Roman culture as the cult of the god Bacchus. The production of beer was so important at that time that it was regulated in the *Code of Hammurabi*, the oldest known legislation in history, which includes the imposition of severe punishments for sellers who did not adhere to the rules of brewing.

However, if any ancient culture is associated with the consumption of beer, it is the Egyptians. It was consumed at any time, and on any occasion. The Egyptians used to drink it mixed with fruits, especially dates, sweetened with honey and perfumed with cinnamon. The passion they had for this drink led them to develope different varieties. Recipes such as strong beer, triple beer, sweet beer, thick beer, coagulated beer, and friendship beer have survived to the present day. They also had access to some imported "blondes." Various illustrated documentation of the time (e.g., murals), show that the first step in beer making consisted of creating a yeast-rich cereal dough, which was lightly baked in the form of loaves. Once this process was finished, the bread was crumbled in water; it is possible that at this point the additives necessary to flavor the different types of beer were added. The thick mixture was left to ferment, and when the process was finished, the liquid was put into its corresponding containers. The result was a beer with a low alcohol content, which in its less refined and more common version was quite thick—more similar to porridge than the sparkling liquid we drink today. At first, the Egyptians obtained beer by fermenting wheat, but later, this cereal was replaced by other more suitable ones, especially barley, which was grown in the Nile delta. It was precisely this barley wine that eventually became identified with Egyptian civilization. The association between beer and the Egyptians was so deep that the Greek and Roman civilizations kept the term *zythum,* which was the word used by the Egyptians to refer to this drink.

Several centuries later, in the Middle Ages, when Christian Europe was facing the Vikings, beer brewing spread throughout the Mediterranean. Monasteries began to add hops, giving beer the particular bitter taste of the current drink, and which also acts as a preservative due to its antimicrobial activity. The brewing techniques were perfected, and the first craft breweries were set up in the tenth century. The acceptance of beer was such that the Magna Carta of England in 1215 (a medieval document of 63 articles on the liberties, privileges, and rights of the nobles, considered by many to be the predecessor of English parliamentarism) guaranteed a standard "...measure for ale..." throughout the realm. In Germany, approximately five hundred cloisters commercialized it, and in 1516, the *Reinheitsgebot*, also known as the German Beer Purity Law or Bavarian Purity Law, was promulgated to regulate its production.

However, it was not until the invention of the microscope in 1837 that yeasts were linked to beer. Three researchers, Charles Cagniard-Latour, Theodor Schwann and Friedrich Kützing simultaneously but independently detected the presence of microorganisms in beer foam using the microscope. The microorganisms observed were yeasts in the process of reproduction, which indicated that they were alive, and they related them to the production of beer. They named the yeast *Saccharomyces*, the sugar fungus, for its ability to convert sugar into alcohol. Later, industrial progress brought improvements in fermentation processes, more efficient machines, and the appearance of numerous commercial breweries.

The production of beer has remained basically the same, but the processes have been greatly refined to improve its production and the final product. The process begins by grinding the barley grains, which will subsequently favor the germination and activation of the enzymes responsible for breaking down the carbohydrates. Next, the barley grains are moistened at a suitable temperature, which causes germination to begin. This stage of the process is known as *malting* or *saccharification*. The water is then removed from the malt to stop the reactions that occur in the grains. This is followed by a roasting process, in which hot air is passed through the grains. Malting and roasting are two stages that must be very well- controlled because they affect the body, color, flavor, and smell of the final product. The process continues with a mashing of the grains in water, then a filtration to separate the husks from the liquid, now called wort, which has a sweet taste. This barley wort is then boiled again, and hops are added. Currently, liquid extracts or tablets are used to achieve a better yield and greater stability of the product. After boiling and the addition of hops, the yeast is now in an environment full of sugar and water—an ideal place to grow; without oxygen, the yeast initiates the process of fermentation in order to consume all that sugar, producing ethanol and $CO_2$ in the process. Thus far we have spoken of a single type of yeast, yet it is possible to imagine that with the biodiversity that exists on the planet there is more than one type of yeast that can be used to produce beer. In fact, according to the type of yeast used, there are two main types of beer: *Lager*-type beers, related to *Saccharomyces carlsbergensis*, which settle to the bottom of the fermentation tank; and *ale*-type beers, with surface yeasts that float in the fermentation tank, such as *Saccharomyces cerevisiae*. This also affects the characteristics of the final product, such as color and alcohol content. Beer and bread are two of the best examples of how fungi have accompanied us in our development and how we have improved their production thanks to the advantages they offer us in our diet.

# Winemaking

Another very important drink in ancient cultures was wine. We can find documents and hieroglyphs showing that this drink was commonly consumed by Egyptian nobles. However, if we have to link wine to a culture, we will inevitably turn to the Greek and Roman cultures, which intimately associated the consumption of wine and celebration. It is no coincidence that the Greek god of the party, Dionysus, or Bacchus for the Romans, was always represented with a chalice full of wine in his hand. In his honor, people went to the countryside to get drunk in an annual party known as *the bacchanals*, which have survived to this day as the *Carnivals.*

The production of wine is easier than that of beer because grapes have a large amount of free sugars, which makes the process easier. To produce wine, the grapes are pressed to obtain the *must* (freshly squeezed grape juice), which contains the sugars, and then immediately left to ferment in a vat. The microbiological aspects of winemaking are very important; wines can be made using the microorganisms naturally present on grape skins, but the mixture of bacteria and yeasts living on grape skins has unpredictable results. To avoid this problem, fresh must can be heated to eliminate these unknown microorganisms, and then the desired strain added. Today, pure and selected cultures are added to ensure the fermentation process is completed at the desired time, and to avoid the invasion of other fungi. As in beer, the yeast most commonly used to make wine is *S. cerevisiae.* The best wine varieties are *S. ellipsoideus* and *S. pastorianus,* which differ from the former mainly because they can continue fermenting up to alcohol concentrations of 18% (vol/vol) and are more resistant to the presence of acids, tannins and sulphur dioxide, allowing the latter compound to be used to inhibit the growth of other organisms. Most of the fermentation takes place in the first 5–7 days, after which there is a decrease in the production of carbon dioxide. In wineries where the carbon dioxide is released into the facility, it is dangerous to remain in the area where fermentation takes place during the period of increased carbon dioxide formation; as the level of oxygen decreases problems of asphyxiation can occur. The newly formed wine, which is called *young wine*, must be protected from contact with the air to avoid developing a *veil* (growth of acetic bacteria). These bacteria oxidize the ethanol, turning it into vinegar and ruining the brew.

Sometimes different types of yeasts and other microorganisms are used to obtain different types of wine, but the most important difference, the color of the wine (red or white), does not depend on the yeast but on the grape used and the winemaking process. From white grapes, only white wine can be obtained, but from black grapes, red wine or white wine can be obtained. This is because red coloring is found in the skin of the grape. If the skin is left in the must that is fermenting, the wine will be dark in color because the alcohol extracts the dye. However, the taste and aroma of wine are generated by a complex mixture of molecules. Some of these compounds are produced during fermentation, and others come directly from the fruit.

## Preparation of Soy Sauce

Thus far we have talked about foods such as cereals and fruits, but let us not forget that legumes also have carbohydrates and therefore can also be fermented. Soya is a legume with a multitude of uses in Eastern societies such as Japanese, Chinese and Indonesian. One of the best-known products is soy sauce, but depending on how the soybeans are processed, other products such as *miso* can be obtained. The process for creating these products is the same. First, the soybeans are boiled and moistened. This is followed by the addition of *koji*, an enzyme preparation grown on cereals or sometimes legumes, and used as a starter for the production of large quantities of fermented foods. In the case of soy sauce, koji seed (*tane*) is produced by cultivating single or mixed strains of the fungi *Aspergillus oryzae* or *Aspergillus soja*. The inoculation of the koji is the determining moment when one product or the other is obtained. If the soybeans are immersed in brine (water to which a large amount of salt has been added) and left for four months of stirring, we will obtain soy sauce. This process is slow and must be very well controlled to avoid the appearance of other organisms that affect the taste of the final product. If, on the other hand, koji is added directly to the soybeans after they have been cooked and left to ferment, *miso* is obtained. Here, the cooked and cooled soybeans are mixed with salted koji and an inoculum consisting of a portion of miso from a previous batch, or pure cultures of yeast and osmophilic bacteria. The mixture, known as green miso, is transferred to tanks or vats where it will undergo fermentation and maturation.

## Cocoa Processing

Cocoa is another food in which the development of its aroma and flavor is based on a fermentation process. Commercial cocoa is derived from the seeds of the ripe fruit of the *Theobroma cacao* plant, native to the Amazon region of South America. Again, the origin of this process is lost in antiquity, but it is believed that in the past, fermentation was carried out simply to help remove the mucilaginous, slimy pulp surrounding the seed for easier drying and storage. Cocoa was introduced to Europe in the fifteenth century by Cortés during the period of the discovery and colonization of the American continent. Its popular demand led to the establishment and dispersal of the plant to virtually all European colonies located between 15 degrees north and south of the equator, where climates allow the cultivation of cocoa. Today, the main reason for fermenting cocoa is to induce transformations in the seeds that lead to the creation of precursors for the aroma, flavor and color of chocolate. Without this treatment, the cocoa beans would be excessively bitter and astringent and, when processed, would not develop the distinctive aroma and flavor of chocolate. The character and intensity of these features are governed primarily by the genetic constitution of the cocoa variety, while fermentation unlocks its full promise. The inherited characteristics of the seed therefore determine the potential that can be reached by fermentation. While it is impossible to improve genetically inferior material with better processing techniques, it is quite easy to ruin good quality cocoa with careless or inadequate curing.

Of the different species of *Theobroma*, only *T. cacao* produces pods, the fruits suitable for making chocolate. Fermentation begins immediately after the fruit is removed from the tree because a number of microorganisms are inoculated from the surface of the fruit, knives, the hands of workers, the containers used to transport the seeds to the fermentation plant, and the dried mucilage on the surface of the fermentation boxes. The cocoa fruit is divided into two parts: the pulp and the seeds. It is the pulp surrounding the seeds that undergoes microbial fermentation, while it is the seeds that are used to make cocoa. Chemical changes that occur within the seed, due to the fermentation of the pulp, make them suitable for processing and consumption. Initially, the *testa* (outer layer) of the seed functions as a natural barrier between the microbial fermentative activities on the outside of the seed and the chemical reactions inside the seed. However, ethanol, acetic acid, and water of microbial origin penetrate the seed, killing the seed. Once it dies, some of its components are filtered through the skin and drained. The death of the seed is a critical event during the fermentation of cocoa, which allows biochemical reactions responsible for the development of aroma and flavor to occur within the seed. Unlike the foods mentioned above, in this case, several yeasts are involved in the production of cocoa. Not all yeasts that contribute to fermentation are present simultaneously, but follow in succession. The successor yeasts feed on the waste produced by their predecessors.

## Fungi as a Source of Antibiotics

Although mushrooms have been with us since the beginning of humankind, our relationship with them did not become serious until approximately 90 years ago. A serendipitous discovery changed the way we understand medicine and changed our life expectancy, increasing it to the values we know today. This began in the summer of 1928, when the British bacteriologist Alexander Fleming discovered that the fungus *Penicillium notatum* produced an antibiotic substance that killed bacteria. This substance was consequently given the name *penicillin*. The discovery was due to a combination of coincidences. First, Fleming's colleague worked with fungi in a nearby laboratory, so the spores of these fungi were scattered in the environment. This made it easy for the culture plates to become contaminated during the brief period of bacterial seeding. In addition, Fleming left the culture plates on the laboratory bench instead of on the stove during the two weeks he was away on vacation. This allowed the fungi to grow in an environment designed for bacterial growth. Finding themselves outnumbered by the bacteria, but in an environment conducive to growth, the fungi set in motion all the defense mechanisms at their disposal to survive. They began to synthesize a penicillin that killed the bacteria and allowed the fungi to use those resources to grow. Next, Dr. Fleming did not throw away the plate, assuming it was contaminated, as other researchers had previously done. Driven by his scientific curiosity, he conducted several studies with other bacterial strains demonstrating the ability of fungi to fight bacteria. Fleming stopped studying penicillin because he did not obtain good results and considered that it was not relevant to his studies. However, the study was very striking, and many researchers asked Fleming for samples of his strains to investigate them. Although many of these studies led nowhere, others came to fruition, such as the one carried out by the researchers Ernst Boris Chain and Howard Walter Florey, who increased both the half-life of the product and its potency. This allowed animal studies to be carried out, demonstrating that penicillin was a safe product because at high doses it did not produce toxic and/or secondary effects. They also showed that it was distributed throughout the body through body fluids, and that it was not immediately metabolized in the body.

In 1940, with World War II having already started, clinical trials on humans began. The first test went wrong because the drug had impurities derived from production, so they had to adjust the manufacturing process. This first test was carried out in the United States because Florey travelled there with the intention of stimulating penicillin production. Americans proposed deep-tank fermentation to improve production, which gave positive momentum to industrial production. As a result, Fleming, Florey and Chain were awarded the Nobel Prize for medicine in 1945 for their great discovery.

Penicillin is still used today, along with other fungal antibiotics, such as oleandomycin and cephalosporin, to fight bacterial diseases. However, while the search for new antibiotics has continued over the years, and new antibiotics have been discovered, it has become increasingly more difficult to develop novel antibiotics to add to the existing antibiotic classes. This problem, together with the increasing number of bacterial strains resistant to antimicrobial agents, suggests that in the future, the antibiotic industry will not be able to produce new compounds to replace those that are no longer effective.

## Use of Fungi in Decontamination and Pest Control

As we already know, fungi can destroy wood and organic fabrics such as cotton if they are not properly stored, thus representing a great economic loss. However, we have learned to control the functions that fungi can perform, such as breaking down large and stable molecules, and emitting the toxins these microorganisms produce when feeding. Controlling the production of these toxins is very important because they can have serious effects on the health of human beings, and can produce some types of cancer.

Currently, the ability of fungi to break down large and stable molecules offers a solution for environmental decontamination processes that occurred during the twentieth century, when humankind witnessed great technological developments that completely changed the way we live. Insecticides, for example, have helped to control pests and prevent millions in crop losses. The problem is that insecticides are very stable molecules and they have been accumulating in the soil and infiltrating into aquifers for years, contaminating them; they have even accumulated in the fatty tissues of plants and animals, including humans. All this has caused chronic exposure to poisonous agents that can have severe consequences for the exposed organisms, and for the future of the species, making it necessary to eliminate pesticides from the environment in a safe and economical way., The characteristics of fungi have placed them in the front line of battle. At first blush, it may seem a very radical idea. What can a microscopic organism do against a molecule designed by humans to kill plants, animals, and even fungi? However, has been very successful mainly due to the enormous variety of fungi and their various characteristics—some of which are able to feed on these pesticide molecules.

Petroleum is another substance that seriously pollutes the environment. Throughout recent history, there have been severe environmental accidents involving this substance. Petroleum is composed of molecules similar to the fats and oils that we consume, and that fungi can degrade. It is an organic compound. The problem is that once crude is extracted from oil-rich areas, it is then transported to the rest of the world via huge oil tankers. These tankers sometimes lose their valuable and toxic cargo at sea, polluting hundreds of miles with a black, oily slick incompatible with life. Fungi, with their ability to derive energy from such molecules and survive in all environments, seem the perfect agents to help the planet respond to accidental oil spills at sea. This theory was put to the test in the accident of the Exxon Valdez oil tanker, which ran aground along the coast of Alaska on 24 March 1989, and spilled 37,000 tons of oil, causing serious environmental consequences still being studied today. Because some fungi feed on the compounds that form oil, or produce substances that help to eliminate the remains of crude oil from the environment, the site became a large-scale test of the potential for these fungi to cleanse oil-contaminated land.

The use of fungi for decontamination tasks is increasingly being studied because they have the advantage of requiring moderate monetary investments and low energy consumption, being environmentally safe, and not generating waste equal to or more dangerous than the initial discharge. Therefore, many industries are adopting the use of fungi to reduce their impact on the environment. For example, such as the textile industry is studying the treatment of wastewater with fungi that degrade the molecules used to color fabrics. During the dyeing process, some of the dyes pass into the rinse water, which is then discharged into the rivers that often surround these factories. These compounds seriously affect the balance of the ecosystems and aquifers into which they are discharged, and often have carcinogenic and mutagenic properties that damage the genetic material of cells. In the Aburrá Valley in Antioquia, a region of Colombia where the textile industry is the main economic engine, studies have already been carried out using fungi to eliminate these substances from the environment. Fungi are also being used to treat heavy metals from the mining industry.

Fungi could also have a place in other industrial processes, such as agriculture, especially in pest control; not only do fungi form associations with plants but they also live inside the stems and leaves of plants. In some cases, these fungi produce substances that are unpleasant or even toxic to insects or grazing animals, and can cause health and growth problems in horses, cows, and other animals important in the livestock industry, and which feed on these plants. While they can present dangers, thus protecting lawns and agricultural crops. This protection is so effective that when horses, cows, and other animals important in the livestock industry feed on these plants, they have health and growth problems. They can also present a great opportunity to control rodents, insects, and other fungal pests, so scientists are studying how to harness these properties of fungi to protect lawns and agricultural crops. In fact, researchers are currently investigating fungi for the biological control of pathogens and insect pests, and some of these species are already being used to parasitize insect pests. In addition, pesticides based on fungi are among the safest to use; therefore, they have been proposed as an alternative solution to artificial compounds, as there is a need to find new pesticides that are less harmful to the ecosystem. Countries such as the United States, Russia, Chile, Australia, Brazil, England, Holland, Belgium and Colombia have developed products for the control of insects; the main component of those products is a species of fungus proven to be effective against these insect pests.

Where the ceiling is in the application of these organisms in our society will be determined by generations to come.

## Bibliography

Acuña G (2002) Discovery of penicillin: a milestone in medicine how chance can help the scientist. Rev Med 1:13

Alonso R (1996) La alimentación mediterránea: Historia, cultura, nutrición. vol 93. Icaria Editorial, pp 51–2

Audesirk T, Audesirk G, Byers BE (2003) Nutrition and digestion. In: Quintanar Duarte E (ed) Biology: life on earth. 8th ed. Pearson Education, pp 686–8, 692

Badii MH, Hernández S, Guerrero S (2015) Effect of pesticides on small mammals: implications for sustainability. CULCyT (30)

Carabias J, Meave JA, Valverde T, Cano-Santana Z (2009) Biodiversity. In: Quintanar Duarte E, Hernández Carrasco F (eds) Ecology and environment in the 21st century. 1st ed. Pearson Educación, pp 88–92

Chanagá Vera X, Plácido Escobar J, Marín Montoya M, Pérez Y, del Socorro M (2012) Native fungi with industrial dye degrading potential in the Aburrá Valley, Colombia. Revista Facultad Nacional de Agronomía-Medellín 65(2)

Corona M (2011) History of biotechnology and its applications. Retrieved from: http://siladin.cch

Doyle MP, Beuchat LR, Montville TJ (2001) Food microbiology: fundamentals and frontiers. In: Doyle MP, Beuchat LR, Montville TJ (eds) Food microbiology: fundamentals and frontiers. Zaragoza, Acribia, S.A., pp 655–99

Gaynes R (2017) The discovery of penicillin-new insights after more than 75 years of clinical use. Emerg Infect Dis 23(5):849

Hernández A ed (2003) In: Industrial microbiology. San José, CR EUNED

Ingraham JL, Ingraham CA (1998) Introduction to microbiology. vol 2. Reverté, pp 736–7

Moreno CM, González A, Blanco M (2004) Biological treatments of contaminated soils: hydrocarbon contamination. Fungal applications in bioremediation treatments. Rev Iberoam Micol 21(1):103 20

Motta-Delgado PA, Murcia-Ordoñez B (2011) Entomopathogenic fungi as an alternative for biological pest control. Ambiente & Água-An Interdisciplinary J Appl Sci 6(2)

Ortiz JMP (2003) People of the Nile Valley: Egyptian society during the Pharaonic period. Editorial Complutense, pp 313

Prescott L, Harley J, Klein D. Fungi (eumycota), mucous moulds and aquatic moulds. In: Microbiology. 5th ed. McGraw Hill-Interamericana, pp 597

Prescott L, Harley J, Klein DA (2002) Introduction to microbiology. In: Microbiology. 5th edn. McGraw Hill-Interamericana, pp 11

Prescott L, Harley J, Klein D (2002) Metabolism: release and conservation of energy. In: Mibrobiology. 5th ed. McGraw Hill-Interamericana, pp 185–8

Prescott L, Harley J, Klein D (2002) Microbiology of food. In: Microbiology. 5th ed. McGraw Hill-Interamericana, pp 1044–9

Prescott L, Harley J, Klein D (2002) Microbial nutrition. In: Microbiology. 5th ed. McGraw Hill-Interamericana, pp 100–1

Solomon EP, Berg R, Martin D (2007) Energy and metabolism. In: Biology vol 7. 9th ed. CENGAGE learning, pp 154–61

Solomon EP, Berg LR, Martin DW (2013) How cells produce ATP? energy release pathways. In: Biology. 9th ed. CENGAGE learning, pp 172–89

Solomon EP, Berg LR, Martin DW (2013) The fungi. In: Biology. 9th ed. CENGAGE Learning, pp 615–23

**Antonio González** holds a degree in Food Science and Technology from Universidad Rey Juan Carlos (URJC) and a master's degree in Agricultural Chemistry and Novel Foods—specialty in Novel Food Development—from Universidad Autónoma de Madrid (UAM). He obtained his Ph.D. from the doctoral program in Pain Research at URJC and he is currently an Assistant Professor and Researcher in the Area of Pharmacology, Nutrition and Bromatology of the Department of Basic Health Sciences at Universidad Rey Juan Carlos. He is part of the High-Performance Research Group in Experimental Pharmacology (PHARMAKOM) of Universidad Rey Juan Carlos, and a member of the European Council of Cardiovascular Reasearch (ECCR). His research work focuses on the role of the innate immune system in the cardiovascular alterations derived from both pharmacological treatments and endocrine-metabolic diseases using animal models. Over the past years he has been repeatedly participating in outreach activities such as *Science Week* or the *European Researchers' Night*.

# Obesity-Induced Diabetes Can Cost You an Arm and a Leg... or a Kidney

Patricia Corrales Cordón◉ and Almudena García Carrasco◉

If you often travel by public transportation, whether by train, metro or bus, you may have noticed that your hips often come into contact with the person in the seat next to yours. This has lessto do with the efforts of municipal transport companies to accommodate more passengers by reducing the size of the seats (although it may be true), and more to do with the fact that you are carrying a few extra pounds.

Obesity is an excessive accumulation of body fat that can affect health; therefore, it is considered a disease with numbers that are continuously growing. Obesity is becoming a worldwide epidemic. According to the World Health Organization (WHO), it is one of the main and most worrisome diseases of the twenty-first century in developed countries, and has tripled since 1975. Almost

P. Corrales Cordón (✉) · A. García Carrasco
Area of Biochemistry and Molecular Biology, Department of Basic Health Sciences, Universidad Rey Juan Carlos, Alcorcón, Spain
e-mail: patricia.corrales@urjc.es

A. García Carrasco
e-mail: almudena.gcarrasco@urjc.es

High Performance Research Group in the Study of Molecular Mechanisms of Glucolipotoxicity and Insulin Resistance: Implications in Obesity, Diabetes and Metabolic Syndrome (LIPOBETA), Universidad Rey Juan Carlos, Alcorcón, Spain

P. Corrales Cordón
Consolidated Research Group on Obesity and Type 2 Diabetes: Adipose Tissue Biology (BIOFAT), Universidad Rey Juan Carlos, Alcorcón, Spain

2 billion adults in the world are overweight[1]; 650 million of these are obese. This is twice the population of the USA or almost 1.5 times the population of the European Union. Most alarming, childhood obesity has increased. Forty-one million children under five years of age are considered overweight or obese. That is why experts recommend caution in regard to gaining a few extra pounds. But, how do we know if we are at our ideal weight? If you weigh and measure yourself, you will be able to calculate your body mass index (BMI). This index takes into account not only your weight but also your height, and based on these two parameters, an indicator is constructed. An index equal to or greater than 25 kg/m$^2$ is considered *overweight*—you are a few pounds above your ideal weight; if it is equal to or greater than 30 kg/m$^2$ it is considered *obese*—you are fat.

The causes of weight gain are multiple, but are mainly due to the changes that have occurred in lifestyle and, fundamentally, to low physical activity, coupled with the abandonment of more traditional diets (e.g., the Mediterranean diet) in favor of the well-known "Western diet." Surely you are familiar with the following: start the day with an energy breakfast of delicious croissants, donuts, or pastries (all foods derived from refined flours with a very high amount of sugar), along with an orange juice (if we have not overslept, it will be natural and freshly squeezed), and a glass of milk with cocoa. If you are hungry, you may even eat five times a day so in this case we have the perfect excuse to go to the cafeteria mid-morning for some nuts or energy bars. With all the sugar we will have consumed, at lunchtime we will be low on sugary fuel, so a plate of pasta carbonara will be a great treat (along with some custard for dessert, of course). For an afternoon snack, a piece of fruit would be good, but sugary soft drinks and some chips are more palatable. And when we get home in the evening after a full day's work, who would not prefer a quarter-pound burger to three sad stalks of celery with salt?

Today, we find ourselves in a situation where we are paying the consequences of unhealthy eating habits and a lifestyle with greatly reduced physical activity. Have you ever stopped to think about how your physical activity habits have changed since you were a child? When we were little there were no mobile phones, we went from house to house calling each other to come out to spend some time playing cards, or you would go directly to the park and join your friends who were already there. But, now you can meet on *WhatsApp,* or maybe each friend is in different places, so you drive to their house because using public transport on the weekend is not very appealing. And these are just a few examples. It is increasingly evident that obesity predisposes individuals to metabolic disorders, including cardiovascular disease, dyslipidemia,[2] insulin resistance and diabetes, as well as alterations in kidney function.

---

[1] According to a report issued by the *Food and Agriculture Organization of the United Nations* (FAO) under the title: "The State of Food Security and Nutrition in the World 2018," the world is home to 821 million hungry people.

[2] Dyslipidemia is characterized by high concentrations of lipids, mainly cholesterol and triglycerides. This phenomenon generates an imbalance between the known LDL and HDL, low- and

Insulin resistance is a very common health problem; it refers to a decrease in the ability of insulin produced by the pancreas to carry out its normal physiological functions. This insulin-resistant state usually precedes various situations, such as type 2 *diabetes mellitus* (DM2) or *metabolic syndrome*, which is characterized by a set of pathological alterations that predispose the individual to develop cardiovascular and renal diseases, as well as diabetes. In addition, it is worth mentioning other circumstances, such as pregnancy, where insulin resistance plays a very important role in a physiological way.[3]

How does insulin work? Insulin is a hormone produced by the pancreas, more specifically by specialized cells called pancreatic beta cells, located in the well-known islets of Langerhans and, like any hormone, its function is to regulate the activity of a specific tissue. When insulin is synthesized, it is secreted into the bloodstream, where it is able to reach all the cells of its target tissues to perform its function. In the case of insulin, most of the body's tissues are affected directly or indirectly, as insulin is primarily responsible for controlling the uptake, utilization, and storage of nutrients in the cells. This hormone directly and indirectly affects the function of most tissues and organs. Its most direct effects are rapid, and involve allowing the passage of glucose from the blood into the cellular interior. Glucose is the main currency, or source of energy for the functioning of a cell and, consequently, of a tissue or organ. Think about this: the brain consumes approximately 120 g of glucose per day. That is the equivalent of approximately 15 packets of sugar. But, insulin also has less direct and slower-acting actions, such as stimulating the main glucose storage organs. Glucose is found in free form in the blood. However, not only do we have glucose in the bloodstream, but there are organs such as the liver or muscles, that store glucose in larger polymers, called *glycogen.* If it were not stored in this form, there would be a risk of it leaving those organs, without which glucose would only be available in the blood immediately after eating. Insulin, therefore, stimulates the transformation of these large storage forms that serve as reservoirs of glucose when dietary glucose is not available.

To delve slightly deeper into the action of insulin, it is necessary to know the sequence of action of this hormone. Briefly, the action of insulin begins with the

---

high-density lipoproteins, respectively. Generally, the imbalance tends to be high levels of LDL and/or low levels of HDL.

[3] Metabolically, during gestation there are two distinct stages. During the first two-thirds of gestation there is an anabolic phase, in which the mother increases her body weight due to the accumulation of adipose tissue that serves as an energy reservoir for the rest of the gestation. During the last third of gestation, the maternal metabolism is reversed, and a state of insulin resistance is generated which causes the maternal reserves to be mobilized through the placenta to ensure the correct growth of the fetus. Therefore, the state of insulin resistance at this stage takes place physiologically. The problem arises when a woman begins a pregnancy overweight or obese, which can lead to an exaggerated increase in insulin resistance. This, in turn, triggers a depletion of the capacity of pancreatic cells that secrete the amount of insulin required during pregnancy, causing an exaggerated increase in peripheral glucose. This effect leads to an excessive increase in the size of the fetus, resulting in the development of metabolic and cardiovascular disorders, and fetal malformations, which can even lead to premature neonatal death.

binding of the hormone to its specific receptors, which are located on the cell membrane. These membrane receptors are composed of four pieces (*subunits*), equal two by two: two located on the outside of the cell ($\alpha$); and two that are organized in an extracellular domain ($\beta$), a transmembrane domain and an intracellular domain. The $\alpha$-subunits bind to the $\beta$-subunits in their extracellular domains via disulfide bonds, stabilizing the receptor structure. Insulin binding to the outer $\alpha$-subunit generates a change in the receptor structure that is transmitted to the cytosolic $\beta$-subunit, leading to its activation. It is the same as when your cat climbs onto your lap: no matter how comfortable you are, you must change your entire posture in order to adapt to it. This activation allows the recruitment of adaptor proteins whose purpose is to form molecular aggregates capable of triggering signaling cascades that allow glucose uptake to take place. It is like an online translator: insulin in cellular language would be something like *"TAKE THIS GLUCOSE! TAKE THIS! TAKE THIS!"* or, in more holistic terms, *"SWALLOW!"*.

To understand why insulin resistance occurs, it is necessary to unravel the molecular basis that would explain this phenomenon. After eating a delicious muffin, the pancreas, from its pancreatic beta cells, produces and releases insulin so that the body can remove nutrients from the blood, and incorporate them into our cells. Insulin is able to activate important and evolutionarily conserved metabolic pathways that trigger a series of events capable of promoting the growth, division and survival of cells. To do this, insulin has to bind to its receptor. This receptor is able to provoke a reaction inside the cell, such that glucose from the bloodstream can bind to a glucose transporter protein (GLUT4, *glucose transporter type 4*) to enter into the cell. A glucose transporter protein is located across the lipid membrane that makes up the surface of the cell, and is a transporter *specific* for glucose. When a glucose molecule arrives at the cell surface, in the presence of insulin the transporter opens its gates and allows only that molecule to pass through. Under pathological conditions, GLUT4 does not open its gates despite the presence of insulin, so we would be in a situation of insulin resistance[4] (Fig. 1).

In recent years, numerous studies have reached the same conclusion: most of the metabolic pathologies associated with insulin resistance are established in patients who are overweight and/or obese. Thus, researchers who study this phenomenon focus on the increase in adipose tissue and chronic inflammation, as well as the greater or lesser expandability of adipose tissue during the development of obesity, which are all risk factors for insulin resistance and, possibly, diabetes in the future.

Adipose tissue serves as an energy store, a heat insulator, and a mechanical shock absorber. However, that does not mean it is good; but neither it is all bad.

---

[4] The pathological situation that gives rise to a state of insulin resistance could be due either to the fact that the pancreatic beta cells are destroyed and do not produce insulin (DM1), or because, although insulin is present, the tissues do not respond to it (DM2). This is why type 1 diabetics have traditionally been known as *insulin-dependent*: their bodies do not produce insulin, but they do respond to the insulin they inject. Type 2 diabetics were known as *noninsulin-dependent* because they do produce insulin, but their organs are resistant to its action. This terminology was abandoned as some type 2 diabetics supplement their treatment with insulin injections.

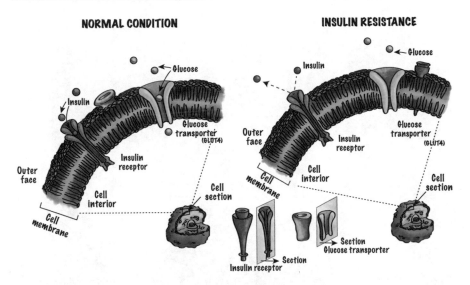

**Fig. 1** **Insulin, glucose transporter and its relationship to insulin resistance.** On the left is a normal situation in which insulin binds to its receptor, triggering a cascade of reactions such that glucose can enter into the cell. On the right, insulin is not able to generate the reactions necessary for glucose to enter the cell; therefore, the cell is not able to "swallow" glucose and would be in a pathological situation of insulin resistance

It is formed before birth and continues to develop throughout life. It accumulates preferentially in the subcutaneous fatty tissue, and is distributed on the sides of the waist in women, and abdominally in men (manifesting as a balloon-shaped belly), the latter distribution being associated with an increased cardiovascular risk. Traditionally, adipose tissue was believed to be a very inactive tissue and that it was simply responsible for forming the aforementioned accumulations in subcutaneous fatty tissue; however, currently, it is known to actually be another endocrine organ, as it establishes constant communication with different organs through the release of *adipokines* (hormones synthesized and secreted exclusively by adipose tissue), signaling proteins synthesized by itself and that, as we will see later, play a key role in the effects of obesity.

Under normal circumstances, excess calories are stored efficiently as adipose tissue cells (*adipocytes*), and provide a reserve of dietary lipids, causing both the size and number of these cells to increase. This process is known as *adipose tissue expandability*. However, when the storage capacity of adipose tissue is exceeded or when fat accumulation occurs at a rate too fast for proper lipid storage, the excess lipids can be burned and/or accumulate in other organs that are not designed to accumulate fat, such as the liver, skeletal muscle, pancreas, and even the kidney; yes, fat can also be stored in the kidney. This ectopic fat accumulation is known as *lipotoxicity* and is one of the main known causes of the development of insulin resistance (Fig. 2).

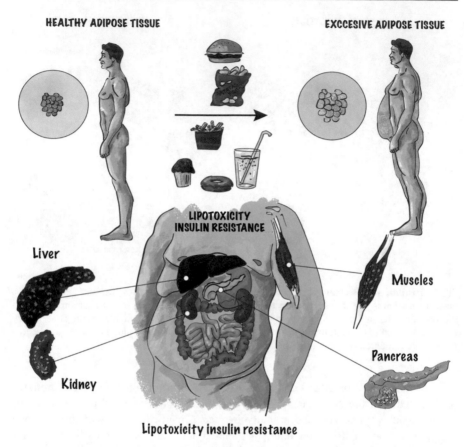

**Fig. 2  Consequences of obesity on adipose tissue: lipotoxicity and insulin resistance.** Failure of adipose tissue expandability in obesity can lead to fat storage in other tissues, such as the liver, skeletal muscle, pancreas and kidney. All these effects will eventually lead to insulin resistance in the body, the beginning of diabetes

## What is a Kidney for?

Have you seen how baleen whales feed? They swallow a huge mouthful of water containing their food, such as krill, and expel it again. They have baleen plates that filter the food, retaining what is of interest to them and eliminating the excess water. The kidney has a similar function: it filters 190 L of blood a day to eliminate approximately 2 L of waste products and excess water, which are converted into urine and stored in the bladder until they are eliminated by urination—a very quiet cleaning robot. If these waste substances and excess water were not eliminated, they would accumulate in the blood, seriously damaging the body.

How does blood processing actually take place in the kidney? Waste disposal takes place in the anatomical and functional units of the kidney and nephrons. Each nephron is composed of, among other things, *glomeruli*, the structures where blood filtration takes place. These glomeruli are made up of blood capillaries folded into a ball, responsible for conducting blood out of and into the glomerulus. In turn, embracing the capillaries, we find the *podocytes*, the cells responsible for forming a filtration barrier as they emit *cellular projections* that intertwine with those of neighboring podocytes, forming a filter that allows the selective passage of some molecules. This ball of capillaries, together with the attached podocytes, is, in turn, surrounded by *Bowman's capsule*, which acts as a funnel to collect the first filtrate that is generated and composed of waste, excess water, and some molecules that will be reabsorbed later. Proteins and cells do not pass through the filter generated by the podocytes but are retained in the bloodstream (water and smaller candies have managed to escape from your bottle, but not the larger ones) (Fig. 3).

Once the blood has finished going through the glomerular capillaries, it will leave the glomerulus through the *efferent arteriole*. All the particles that have managed to cross the filtration barrier will be conducted from Bowman's space towards the renal tubules, where the reabsorption of chemical compounds such as glucose, sodium, phosphorus, and potassium will continue to take place. Finally, waste substances and excess water remaining in the proximal and distal tubules pass through the collecting tubule and flow into the ureters to reach the bladder, where they will be eliminated. A whole network of processes is generated with the ultimate goal of eliminating all the waste substances that the body is not able to store, and that are detrimental to its proper functioning.

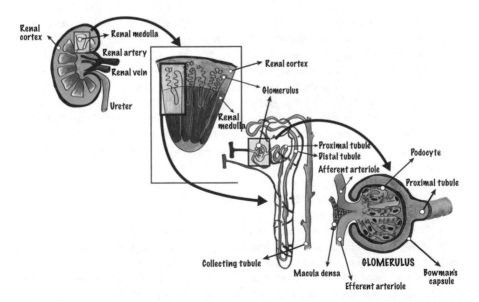

**Fig. 3  Structure of the kidney.** From left to right, the structure of the kidney is shown from the simplest to the most complex structure. On the right, the details of the structure of a glomerulus, the structure where blood filtration occurs, are not missed

## Obesity Damages the Kidneys

Obesity causes adipose tissue to expand, and its metabolic needs to increase. If we do not stop eating and storing excessive amounts of sugar in our body, in an attempt to keep this growing tissue properly irrigated, compensatory mechanisms are put in place to ensure both its functioning and that of the rest of the individual. The mass of some organs, such as the kidneys, increases, as well as the work to be done by the heart to cope with the increased blood that occurs. In other words, a larger individual needs more blood, a heart that pumps more, and larger organs. It is not surprising that this increase in the demand and function of the body leads to wear and tear and to a more rapid loss of kidney function than in healthy individuals.

Thus, the volume of fluid that has leaked into the glomeruli is greater. This generates a state of *glomerular hypertension*, observable in obese patients whether they have general hypertension or not. However, it is not that simple; the blood flow of the kidneys can also be altered by the accumulation of fat deposited in the *renal sinuses*, and these accumulations can compress the main vessels supplying the kidney (renal artery and vein). This leads to a reduction in oxygen supply to the tissue (hypoxia conditions), which only further complicates the pathology.

As you have already learned in this chapter, obesity is associated with insulin resistance. In the case of the kidneys, it has been linked to sodium retention and increased filtration rate in the glomerulus. It may also influence podocyte morphology, remodeling, survival, and function.

One of the most damaging effects of obesity is the alteration of the *renin–angiotensin–aldosterone system*, or RAAS, (a somewhat convoluted name, but ideal for looking like a real expert at Christmas dinners in front of your "know-it-all" brother-in-law). This system is responsible for regulating blood volume and blood pressure and is largely globalized because none of its components remain in the organ generated but goes out into the world. Under normal conditions, when renal flow decreases, the kidneys release *renin*, which travels to the liver, where it is converted from angiotensinogen to angiotensin I. This, in turn, travels to the lungs, where it will be converted into angiotensin II by the action of the aptly-named enzyme: angiotensin converting enzyme (ACE). Angiotensin II is a potent vasoconstrictor that leads to an increase in blood pressure. When angiotensin II reaches the adrenal cortex (the outermost part of the adrenal glands, located above the kidneys), aldosterone is released. This aldosterone is responsible for increasing tubular reabsorption of sodium and water, leading to an increase in extracellular body fluids and, therefore, an increase in blood pressure (Fig. 4).

In obesity the *renin–angiotensin–aldosterone system* is negatively affected (how could it be otherwise?). It has been observed that adipose tissue can activate it in two ways: first, through the release of angiotensinogen; and second, through the release of mediator molecules that activate the adrenal gland and, with it, the release of aldosterone (Fig. 4). In addition, circulating levels of renin are increased, which is a risk factor for renal damage. Furthermore, obesity also causes

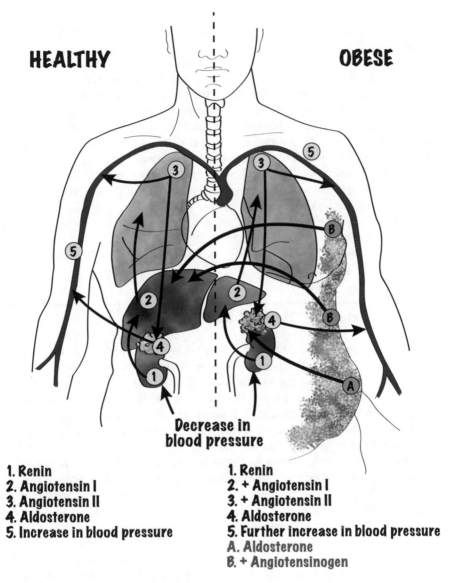

**HEALTHY**

**OBESE**

**Decrease in blood pressure**

1. Renin
2. Angiotensin I
3. Angiotensin II
4. Aldosterone
5. Increase in blood pressure

1. Renin
2. + Angiotensin I
3. + Angiotensin II
4. Aldosterone
5. Further increase in blood pressure
A. Aldosterone
B. + Angiotensinogen

**Fig. 4  Renin–angiotensin–aldosterone system.** On the left you will find a healthy situation in which the renin–angiotensin–aldosterone system is working correctly. On the right, you will find the effects of this complex system when the individual is obese, all of them alterations generated by the large accumulation of fat in the body

an increase in *angiotensin II synthesis*, leading to a constriction of the glomerular efferent arteriole, together with the development of fibrosis and abundant podocyte *apoptosis*.

The high activation of the renin–angiotensin–aldosterone system in obese individuals leads to increased sodium reabsorption in the proximal tubule, thus reducing the sodium supply to the distal tubule. Therefore, a feedback mechanism based on communication (due to their anatomical proximity) between the cells of the *macula densa* and the glomerulus, known as *tubulo-glomerular feedback*, is set in motion. In this process, the cells of the macula densa are able to detect changes in the sodium concentration in the distal tubule (as if testing the salt level of the filtered product), and transmit the information to the glomerulus; the tone of the afferent arteriole is regulated, and with it, the blood flows to the glomerulus. As a consequence of the decrease in sodium levels detected by the macula densa, vasodilatation of the afferent arteriole is generated, which, together with the vasoconstriction of the efferent arteriole mentioned above, leads to an increase in the pressure inside the glomerulus and, therefore, to the hyperfiltration characteristic of obesity.

## Consequences of an Obese Kidney

With all these negative factors, the detrimental effects of obesity on the kidneys are not long in coming. The main consequence of obesity in the kidneys is the accumulation of fat inside the kidney itself. This causes an increase in the size of the organ, and increases the pressure inside the kidney, changing the flow of renal fluids and altering glomerular filtration. The accumulation of lipids inside renal cells, especially in podocytes, leads to an alteration of their functions, loss of structure, fibrosis, and even cell apoptosis.

An important effect of obesity on the kidney is the associated glomerulopathy, the main features of which are glomerulomegaly[5] and secondary focal glomerulosclerosis.[6] As the size of the glomerulus increases, the podocytes are forced to increase in size and extend their processes to compensate for the areas that have been left bare, and thus maintain the integrity of the filtration barrier. However, this effort by the podocytes is not sufficient to maintain the filtration barrier. Since these cells cannot multiply, and their ability to expand does not correlate with the increase in size of the glomerulus, they are subject to forces of tension and stress. This situation will reach a critical point for the podocytes, which will eventually detach and fall into the urine (Fig. 5).

In the glomerulus, the progressive loss of podocytes leaves locally bare areas on the capillary ball. This results in a loss of the selectivity of the filtration membrane:

---

[5] Increase in the size of the glomerulus.

[6] The term *focal segmental glomerulosclerosis* refers to the appearance of collagen fibers in some glomeruli as a consequence of damage due to obesity.

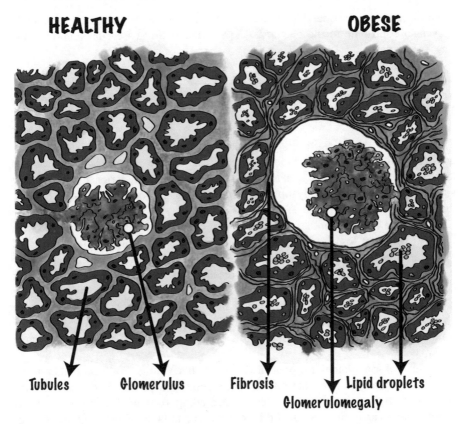

**HEALTHY**                                    **OBESE**

Tubules          Glomerulus          Fibrosis          Lipid droplets

Glomerulomegaly

**Fig. 5 Fat accumulation, glomerulomegaly and/or fibrosis in the kidney.** On the right are features of healthy kidney tissue. On the left is a glomerulus with large size, collagen fibers between the tubules demonstrating fibrosis, and accumulation of lipid droplets, all characteristic of a cross-sectional image of a kidney from an obese individual

proteins that were previously retained in the blood can now pass into the urine[7] (large candies can now escape from the bottle). The spaces left by the podocytes will be treated like any other wound that occurs in the body: a scar will form based on the synthesis of collagen fibers in the glomerulus (*glomerulosclerosis*). In this process, highly differentiated cells are replaced by collagen fibers that are not able to fulfil the function of the podocytes, so a nonfunctional fibrotic organ will be generated, which will ultimately need to be replaced.

Clinically, glomerulomegaly may not manifest renal symptoms. The availability of histological samples from patients requires a biopsy, so it is not always possible to identify this alteration. This reduces the diagnosis of this pathology to the presence of proteinuria that is not always detected in time and can lead to end-stage

---

[7] The presence of high levels of protein in the blood is called *proteinuria*.

renal disease in a considerable number of patients. Hypertension and dyslipidemia may also contribute to the diagnosis. As in many other cases, prevention is the best medicine.

## Treatment for the Improvement of Obesity-Associated Kidney Damage

The two main strategies in the treatment of obesity-associated glomerulomegaly are weight loss and RAAS blockade, but it should be noted that all measures aimed at improving obesity-related risk factors (such as hypertension, metabolic syndrome, diabetes, and dyslipidemia) contribute to the improvement of the disease.

One of the current frontiers in the treatment of glomerulomegaly associated with obesity is weight loss by controlling caloric intake, or through bariatric surgery. When an obese person loses weight, the effects can be observed in a short time: decreased proteinuria, improved blood pressure, lower accumulation of lipids in the blood, improved circulating glucose levels, and insulin sensitivity. The main effect on the kidney is the improvement in the filtration rate. This effect is not observed, however, with liposuction, the alternative chosen by some when they want to lose weight; a negative intake of calories seems necessary to achieve such benefits.

Not all calorie reduction approaches have positive results. For example, bariatric surgery modifies the intestinal microbiota; a reduction of the *Oxalobacter formigenes* bacteria, an organism capable of degrading oxalate, has been observed, which leads to an accumulation of this compound and favors the appearance of kidney stones. Due to the importance of the intestinal microbiota in the inflammation associated with obesity and other complications, fecal transplants have recently been postulated as an alternative to improve it.

Blockade of the RAAS by means of angiotensin-converting enzyme inhibitors or angiotensin II receptor blockers is the second pillar on which the treatment of obesity-associated glomerulomegaly is based. It has been observed that obese patients show a greater sensitivity to the renoprotective effects of RAAS blocking drugs than non-obese patients, and through this treatment achieve a reduction in proteinuria of up to 80%.

## Towards a New Horizon

Obesity is a global epidemiological problem that continues to grow at alarming levels—with no signs of slowing down. This has major health consequences due to the association of obesity with the development of other cardiovascular diseases, cancer, insulin resistance and DT2, as well as alterations in other organs such as the kidneys.

Once adipose tissue has reached and outgrown its maximum expansion point, it is unable to continue storing fat, and lipids escape from it to other organs not specialized to deal with them. Among the lipid receptor organs are the kidneys. The accumulation of fat in the kidneys is responsible for the modification of renal blood flow and the alteration of the metabolic functions of the podocytes, the cells responsible for blood filtration.

In this sense, efforts aimed at improving the obesogenic and overweight status of individuals, as well as the impact of the development of other associated diseases, should be priority objectives in clinical and basic research on obesity. As researchers, we have already rolled up our sleeves, and set out to combat obesity and its effects on the body. We look forward to telling you more in the not too distant future. In the meantime, remember: eat more vegetables, strap on your smart watch and go for a walk—now that getting your steps in is all the rage.

## Bibliography

Bosma RJ, Krikken JA, Homan van der Heide JJ, de Jong PE, Navis GJ (2006) Obesity and renal hemodynamics. Contrib to Nephrol 151:184–202

Câmara NO, Iseki K, Kramer H, Liu ZH, Sharma K (2017) Kidney disease and obesity: epidemiology, mechanisms and treatment. Nat Rev Nephrol 13(3):181–190

Carobbio S, Pellegrinelli V, Vidal-Puig A (2017) Adipose tissue function and expandability as determinants of lipotoxicity and the metabolic syndrome. Adv Exp Med Biol 960:161–196

Corrales P, Vidal-Puig A, Medina-Gómez G (2018) PPARs and metabolic disorders associated with challenged adipose tissue plasticity. Int J Mol Sci 19(7):e2124

García-Carrasco A, Izquierdo-Lahuerta A, Medina-Gómez G (2021) The kidney-heart connection in obesity. Nephron 145(6):604–608

Kanasaki K, Kitada M, Kanasaki M, Koya D (2013) The biological consequence of obesity on the kidney. Nephrol Dial Transplant 28(Suppl 4):iv1–7

Medina-Gomez G, Gray SL, Yetukuri L, Shimomura K, Virtue S, Campbell M, Curtis RK, Jimenez-Linan M, Blount M, Yeo GS, Lopez M, Seppänen-Laakso T, Ashcroft FM, Oresic M, Vidal-Puig A (2007) PPAR gamma 2 prevents lipotoxicity by controlling adipose tissue expandability and peripheral lipid metabolism. PLoS Genet 3(4):e64

Naumnik B, Myśliwiec M (2010) Renal consequences of obesity. Med Sci Monit 16(8):RA163–70

Redon J, Lurbe E (2015) The kidney in obesity. Curr Hypertens Rep 17(6):555

Rodríguez-Rodríguez E, Perea JM, López-Sobaler AM, Ortega RM (2009) Obesity, insulin resistance and increase in adipokines levels: importance of the diet and physical activity. Nutr Hosp 24(4):415–421

Virtue S, Vidal-Puig A (2010) Adipose tissue expandability, lipotoxicity and the metabolic syndrome–an allostatic perspective. Biochim Biophys Acta 1801(3):338–349

**Patricia Corrales Cordón** has a degree in Biology and a master's degree in Molecular and Cellular Biology from Universidad Autónoma de Madrid (UAM), and obtained her Ph.D. from the Doctoral Program in Health Sciences at Universidad Rey Juan Carlos (URJC). She is currently Assistant Professor and Researcher in the Area of Biochemistry and Molecular Genetics of the Department of Basic Health Sciences at Universidad Rey Juan Carlos. She makes part of the High Performance Research Group in the Study of the Molecular Mechanisms of Glycolipotoxicity and Insulin Resistance: Implications in Obesity, Diabetes and Metabolic Syndrome (LipoBeta), and of the Consolidated Research Group in Obesity and Type 2 Diabetes: Adipose Tissue Biology (BIO-FAT) of Universidad Rey Juan Carlos. She is a member of the Spanish Society of Biochemistry and Molecular Biology (SEBBM), the Spanish Society of Endocrinology and Nutrition (SEEN), the Spanish Society of Obesity (SEEDO) and the European Society for the Study of Diabetes (EASD). Her research work focuses on the study of metabolic diseases such as obesity and diabetes, particularly in the modulation of adipose tissue both in animal models and in humans. As a disseminator, she has experience as a speaker at science festivals (e.g., *Pint of Science*) and outreach activities (e.g., *Science Week, the European Researchers' Night* and the SEBBM *Teacher's Corner*).

**Almudena García Carrasco** has a degree in Pharmacy from Universidad Complutense de Madrid (UCM) and a master's degree in Biomolecules and Cell Dynamics from Universidad Autónoma de Madrid (UAM). She is currently a researcher in the Biochemistry Area of the Department of Basic Health Sciences at Universidad Rey Juan Carlos (URJC). She belongs to the High-Performance Group in the Study of the Molecular Mechanisms of Glycolipotoxicity and Insulin Resistance: Implications in Obesity, Diabetes and Metabolic Syndrome (LipoBeta) of Universidad Rey Juan Carlos, and is a member of the European Atherosclerosis Society (EAS) and the European Renal Association-European Dialysis and Transplantation Association (ERA-EDTA). Her research work focuses on the study of the role of nuclear receptors in renal podocytes in metabolic diseases, both in animal models and in humans. As a disseminator, she collaborates with the gazette of *Asociación Cultural Aula Altamira*, a nonprofit association dedicated to literacy and adult education, by producing a series of short stories. She also has experience as a speaker at science festivals (e.g., *Pint of Science*) and outreach activities (e.g., *Science Week*).

# The Science of Living Longer

Elena López Guadamillas◉

Attaining immortality or finding the fountain of eternal youth have been recurring themes and a constant quest of mankind since time immemorial. According to the book of Genesis, Methuselah must have been pretty close because it states that he lived to be 969 years old; hence the saying "older than Methuselah." Beliefs aside, what we can be sure of is that the French woman Jeanne Louise Calment lived to 122 years and 164 days, making her the longest documented person in history (Arles, 21 February 1875—Arles, 4 August 1997). Jeanne, who lived an independent life until the age of 110, survived two world wars and the Great Influenza epidemic of 1918, saw the Eiffel Tower built, and even met Van Gogh when she was 13 years old and he came to her uncle's weaving workshop to buy some canvases. One of her many anecdotes was that in 1965 she signed a contract for the sale of her uncle's flat, for which she would receive a monthly payment and maintain the right to live in it until her death. The notary who bought the house, André-François Raffray, must have thought it was a great deal, as Jeanne was in her 90s at the time. What he could not have expected was that he would end up paying for many years, and his wife would inherit the debt on his death. All told, they ended up paying more than double the real price for the apartment. As Jeanne would say *after the fact*, "in life, sometimes you make bad deals."

Everyone hopes that if they reach that age, they will arrive there in both good physical and mental health. Thanks to scientific advances and improvements in public health, diet, and the environment, among other factors, life expectancy has doubled over the last century. It should be noted that this increase in longevity is largely due to the decrease in infant mortality (especially during the first 4 years of

E. López Guadamillas (✉)
UCL Cancer Institute, University College London, London, UK
e-mail: e.guadamillas@ucl.ac.uk

life), and to the success in the fight against infectious diseases due to the introduction of vaccination programs and the use of antibiotics.[1] To give an example, if in
1910 the average life expectancy at birth in the United States was approximately
50 years (40 years in Europe), a baby born today can expect to live between 75
and 80 years (Source: National Center for Health Statistics and Eurostat).

## Longevity: A Question of Environment or Genetics?

Worldwide, there are so-called *blue zones*: places where there are groups of people
or populations with exceptional longevity, well above average. These blue zones
include Sardinia (Italy), the island of Okinawa (Japan), Loma Linda (California),
the Nicoya Peninsula (Costa Rica), and the island of Icaria (Greece). Interestingly,
they are all subtropical regions with warm, temperate climates. Blue zone inhabitants also share common and characteristic lifestyle patterns that contribute to their
extreme longevity. In general, they follow a reduced calorie diet, rich in vegetables
and legumes, characterized by low meat consumption, and moderate consumption
of alcoholic beverages. Even at an advanced age, they practice a moderate but
regular level of physical activity, have a relaxed philosophy of life, and live in a
social structure very focused on family and personal relationships.

Today, with the exception of these blue zones, one out of every 10,000 people
in industrialized countries enjoys centenarian status. As we have already mentioned, there is no doubt our environment and our lifestyle greatly influence our
life expectancy. What has been discovered in the last 20 years is that genetics (i.e.,
heredity) also plays a very important role in longevity. The first data that pointed
to the fact that our genes could significantly determine life expectancy came from
epidemiological studies carried out on identical twins, who have exactly the same
genes. Based on these studies, it has been shown that the probability of reaching
the age of 90 is determined by genetic inheritance. This means that approximately
25% of longevity seems to be determined by our genes, and to be inheritable.
Its influence seems to be even greater and more significant in cases of extreme
longevity where one is over 100 years old. In other words, if your sibling is already
a centenarian, congratulations: you are 8 times more likely than the rest of the population to also reach the age of 100 if you are a woman, and 17 times more likely
if you are a man.

To demonstrate and explain how our genes influence longevity, a wide variety
of transgenic animals have been generated in the laboratory, making it possible to

---

[1] The first 1–4 years of life are the most critical in human survival. During these years we are still
developing our immune system, so we are more vulnerable to all kinds of infections. After this
threshold, especially in the first year of life, the chances of survival increase drastically. The main
pathologies responsible for these child deaths worldwide include malaria, pneumonia, diarrhea,
premature births and other neonatal infections. To give an idea of the importance that infant deaths
have had on average life expectancy throughout history, it is considered that the control of infant
mortality alone is responsible for 50% of the increase in longevity achieved.

extend their average lifespan (in some cases even more than fivefold) by modifying only one gene. Although many of these genes have already been discovered and described, the number of genes involved in one way or another in longevity may be in the hundreds and have yet to be identified. We, at least for the moment, and much to our regret, can do nothing to change the genetic inheritance that has been handed down to us. But, wait a minute—we just said that it is estimated that only 25% of longevity is heritable. That means that the remaining 75% probability of reaching the select *supercentenarians* club is practically in our own hands! This is precisely where science and knowledge play a key role. That is why scientists are rolling up their sleeves and looking for strategies, from the most basic to the most applied research, so that we can live longer and better.

Let us start with the basics: how do we define aging from a biological point of view, and above all, what causes it? In general, aging is described as the progressive loss of the physiological integrity of the organism due to the accumulation of cellular damage, resulting in the functional decline of the body. As a consequence of the passage of time, our cells and tissues accumulate what we might compare to wounds or scars, and potentially harmful compounds make more mistakes, and ultimately become less efficient. This deterioration means that all organisms lose their physiological capabilities over time and cease to function. It is not surprising, therefore, that age is considered per se the major risk factor for a number of diseases, such as cancer, and neurodegenerative and cardiovascular diseases.[2]

Up to this point, it seems that the scientific community has reached agreement. However, everything becomes more complicated when we try to describe the causes that lead to aging and the molecular characteristics that define it. Although there is some unanimity, science continues to debate and update itself as our knowledge and understanding of the molecular and cellular basis of life increases. A clear example of this constant renewal of ideas is represented by the *free radical theory* or *oxidative theory*, one of the most widespread theories of aging, and probably not unfamiliar to you. According to this theory proposed by Denham Harman in 1965, *eukaryotic cells* (any cell that is not a bacterium) use their mitochondria as energy-producing factories in which waste products are generated as a result of their activity. These waste products are mostly oxidative agents, i.e., reactive oxygen species capable of oxidizing and damaging the cells themselves and their components. Over time, not only do more oxidative species accumulate, but their production increases proportionally as the mitochondria themselves begin to fail. It would seem reasonable to think that these agents would oxidize the cells and their different compartments (just as the parts of an old machine rust), and that this would be one of the reasons why they stop working properly. Despite this, and

---

[2] According to the European Statistical Office (Eurostat), the main causes of death in the European Union are circulatory diseases, followed by cancer, COVID-19 and respiratory diseases. Likewise, the National Center for Health Statistics (NCHS) reports heart disease, cancer, and COVID-19 as the leading causes of death in the USA. On the other hand, if we look at developing countries where basic public health measures have yet to be implemented, the main causes of death are mostly due to respiratory infections, diarrhea, malaria or, complications related to childbirth.

although it has been shown that over time the mitochondria do make more mistakes and generate more reactive species, this theory has been called into question because its validity has not been tested in the laboratory. Contrary to expectations, in yeast (*S. cerevisiae*) and worms (*C. elegans*), two models commonly used to study aging, higher levels of reactive oxygen species increase the longevity of these organisms. Similarly, mice genetically engineered to undergo greater oxidative stress and produce and accumulate more reactive oxygen species do not show accelerated aging. Conversely, mice with enhanced antioxidant defenses do not live significantly longer. In humans, it is not clear whether antioxidant intake is beneficial, rather, it appears to be ineffective and even dangerous in some cases, as several studies have linked antioxidant use to an increased likelihood of certain cancers. Despite having all this scientific evidence on the table, we are constantly bombarded with advertisements for antioxidant products that are, at best, harmless, but at worst, could even be harmful to our health.

## Cells also Age

It is now believed that much of the cellular damage that causes the decline of the organism and accumulates over the years is concentrated mainly in the DNA. We are continually exposed to physical (sun rays, X-rays), chemical (pollution, tobacco) or biological (some viruses) agents that test the integrity and stability of DNA and thus its structure and the genetic information it contains. The regions of DNA most affected by these factors are the *telomeres*, noncoding DNA structures (this means that they do not produce proteins) that are found at both ends of the chromosomes, and whose main function is to protect their structural stability. If we compare chromosomes to shoelaces, telomeres would be the plastic tips on both ends of the shoelace that prevent it from fraying. The problem is that with time and cell divisions, telomeres become shorter and shorter, thus losing the ability to protect the DNA.[3]

---

[3] For a cell to divide, it must first duplicate the genetic information contained in its chromosomes. This is the only way for each of the two resulting daughter cells to contain a complete copy of its DNA. This process, called *replication*, is carried out by the enzyme DNA polymerase, the function of which is to attach itself to the DNA, read its nucleotide sequence (the letters that compose it), and select the pieces to synthesize an identical copy. This enzyme is only able to copy the sequence in one direction. For one of the two strands of DNA this is no problem: the DNA polymerase copies its sequence from start to finish. For the other strand, which runs in the opposite direction, a little help is needed in the form of blocks of RNA. These blocks, known as primers, mark the start of synthesis for a new piece of DNA known as the Okazaki fragment. Subsequently, the primers will be replaced by DNA and the Okazaki fragments linked together. These fragments will be formed consecutively Along the chromosome, but at the end of the DNA, at the telomere, this will not be possible: there is no space to place a primer to mark a new beginning. As a consequence, the telomere becomes shorter and shorter with each replication. The wear and tear and loss of the telomere prevents it from performing its protective function for the chromosome, making it unstable and

Telomeres are important in aging, and their length is considered one of the most reliable and accurate measures of biological age: the shorter the telomere length, the older the individual (Fig. 1). Short telomeres have been associated not only with premature aging, but also with certain diseases such as pulmonary fibrosis and some types of cancers. To counteract the effects of telomere shortening, a single enzyme is capable of lengthening telomeres: telomerase. This is activated spontaneously during embryonic development to allow cell proliferation, but in adult life it allows cell proliferation in many tumor cells, giving them the ability to divide indefinitely—a kind of "cellular immortality." In animal models, it has been shown that lengthening telomeres by activating telomerase reverses aging, improves neuromuscular capacity, prevents bone loss, and significantly increases life expectancy without increasing the incidence of cancer. Taking into account precisely this dual antiaging but protumorogenic function of telomerase, it is essential to continue investigating its properties, and to carry out rigorous clinical trials to clarify its possible use as an antiaging treatment in humans.

Another characteristic shared by aging organisms is the accumulation of senescent cells in tissues. *Cellular senescence* can be defined as the loss of the proliferative capacity of cells. In other words, a cell is capable of dividing a limited number of times until a loss of function, malfunction, or some kind of cellular damage causes its deterioration. In these cases, the cells decide not to grow or divide any more, and enter a static, quiescent state, in which they stop all their growth machinery and much of their activity. As one would expect, the accumulation of these senescent cells increases with age; to a large extent, it is the reason our tissues regenerate less well over time. Although at first glance this may seem a biological blunder, it is actually an essential mechanism for organisms to protect themselves against cancer; by entering into senescence, aged, damaged or malfunctioning cells avoid spreading and, at the same time, send signals to the immune system to be eliminated. Building on this concept, *senolytics* (the use of compounds that selectively kill senescent cells) is considered a possible method to delay or improve aging. This is based on, among others, the fact that the elimination of senescent cells in mice suffering from premature aging delays the onset of age-associated pathologies, and increases their longevity.

## Calorie Restriction as a Key to Living Longer

There is no doubt that much progress has been made in the search for possible strategies to improve our quality of life and extend our longevity. However, and perhaps surprisingly, the single most robust intervention that has allowed us to extend life expectancy in all organisms that have been tested, including yeast, worms, flies, fish, rodents, and even macaques, is calorie restriction. To be clear, we mean a quite considerable reduction in caloric intake (in some cases

more vulnerable. Cells with critically short telomeres are unable to divide, cease to be viable, and activate cell death mechanisms.

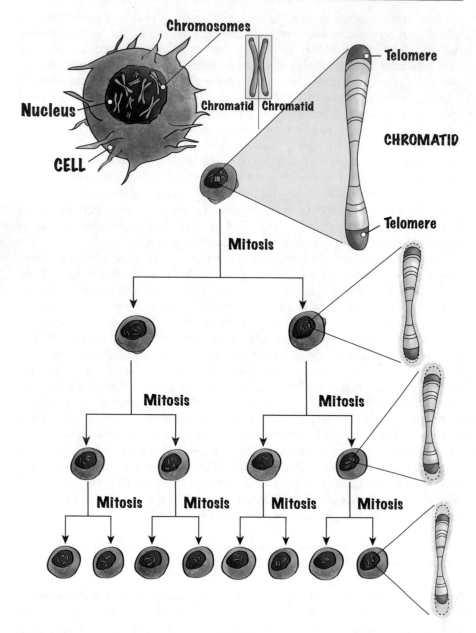

**Fig. 1** **Telomeres as protective elements of DNA**. Telomeres are among the regions most affected by aging, and their length is considered one of the most reliable and accurate measures to establish the biological age of individuals

up to 50%), but without reaching malnutrition, and maintaining a healthy and balanced diet. It is also important to mention that calorie restriction not only prolongs the average lifespan, but also improves the state of health, protecting against cardiovascular diseases, neurodegenerative diseases, diabetes, and cancer.

This finding came about in 1935, when biochemist Clive McCay demonstrated that by restricting the intake of laboratory mice by 25%, it was possible to extend their average lifespan by 15%. This increase in longevity could be further increased as the number of calories was reduced, reaching a maximum extension of 50% in life expectancy with a 55% restriction in the number of calories ingested. Once this threshold was passed, the effects of further reducing intake were counterproductive.

To try to understand why this relationship between calorie restriction and aging is so close and on what it is based, we must once again unravel the molecular bases that explain this phenomenon. After ingesting food, the pancreas produces and releases insulin so that the body can remove nutrients from the blood and incorporate them into our cells. Insulin and other factors such as growth hormone (GH-Growth Hormone) or IGF1 (Insulin-like Growth Factor 1), activate one of the most important and most conserved metabolic pathways in the body: the *insulin-PI3K (Phosphoinositide 3 Kinase) signaling pathway*.[4] Without going into detail, once activated, this pathway triggers a series of events that mainly promote cell growth, division, and survival. It is precisely here, in this "progrowth" response mediated by the insulin-PI3K axis, that the main potential of nutrients in promoting aging is thought to lie, although this remains to be definitively demonstrated. The more a cell grows and divides, the more damage it accumulates, the more it wears out, and the more likely it is to undergo deleterious mutations.

From yeast to humans to mice, flies, and fish, mutations that prevent or reduce the activity of the insulin-PI3K pathway have been shown to significantly increase longevity. Such mutations can occur in several molecular players involved in the PI3K pathway and take place spontaneously in nature or may be artificially introduced in the laboratory. In human populations, we can find several examples that demonstrate the involvement of this metabolic pathway in longevity. First, there is a relationship that seems to be quite clear between height and longevity. One of the first studies was carried out on baseball players who died during the twentieth century, and whose height was reliably recorded; whereas, a similar lifestyle was

---

[4] The insulin-PI3K signaling pathway is one of the main pathways involved in aging. In fact, partial inhibition of its activity protects against cancer, improves insulin sensitivity, and increases longevity in a wide variety of organisms. Their activation occurs when insulin, IGF1 or GH bind to their corresponding receptors located in the cell membrane. Once this extremely specific binding occurs, the phosphokinase PI3K is recruited to the cell membrane and activated. As a consequence, PI3K performs its function as a phosphokinase by *phosphorylating* (adding a phosphate group) the phospholipid PI(4,5)P2 to generate PI(3,4,5)P3. PI(3,4,5)P3 in turn acts as a secondary messenger that transmits the activation signal to a group of proteins containing a PH (pleckstrin homology) domain including AKT, mTORC2 and PDK1. It is these latter members that promote growth, survival, or division responses in cells.

*assumed* in all of them. The result obtained shows an almost linear relationship: the greater the height is, the shorter the life expectancy. Another example is provided by individuals belonging to an isolated community in Ecuador who suffer from growth hormone (GH) deficiencies from birth. They tend to be considerably shorter than normal, too, but what has attracted attention is that diseases such as diabetes or cancer are very rare in them; and although these individuals show a high mortality at an early age for reasons associated with lack of resources, social exclusion, or accidental deaths, it seems that they also have a greater longevity.

As you can guess, this first link established between longevity and nutrition by McCay opened up endless lines of research and possibilities to study how to delay aging through diet. While calorie restriction works in all the organisms in which it has been tested, including mammals such as rats, mice and monkeys, there are very significant differences between species, subspecies, sexes, age of implantation. For example, while in certain strains of mice reducing intake by 50% results in the maximum possible increase in longevity (reducing beyond 50% is harmful), in others this limit is reached with a decrease of less than 20% and, in the case of rats, with only 10%. Similarly, female mice benefit much more from calorie restriction than males in terms of longevity. Another important conclusion reached is that calorie restriction is not all positive. Despite reducing the risk of certain diseases such as diabetes, cancer or cardiovascular disease, calorie restriction can also be dangerous because it weakens the immune system, leaving organisms defenseless against possible infections, reduces fertility, and worsens cell regeneration.

One of the first challenges that scientists faced after these observations was to determine whether the reduction of any particular nutrient was essential to achieve the beneficial effects of calorie restriction. Could longevity be extended by reducing the consumption of carbohydrates, proteins, or fats exclusively? Much progress has been made in this regard, although there is still considerable controversy and different points of view. A number of experiments in flies and mice suggest that it is indeed the balance of macronutrients in the diet and not, as previously thought, the total caloric intake alone that explains much of this effect. They show that protein restriction, and more specifically the amino acids tryptophan and methionine,[5] promote improvements in health similar to those obtained with calorie restriction. However, in recent years, the importance of ketogenic diets,[6] which are low in

---

[5] Amino acids are the organic units that make up all proteins. There are 20 different amino acids, of which 9 are called "essential." Unlike the nonessential ones, the essential amino acids are those that, being essential, the body itself is not able to synthesize, so they must be ingested with the diet. In this group we find, among others, tryptophan and methionine.

[6] Ketogenic diets are those diets characterized by a high content of fat and protein, and a low percentage of carbohydrates (glucose), and usually prescribed mainly as a treatment for childhood epilepsy. Its name alludes to its ability to generate a response similar to fasting known as ketosis, in which, in the absence of glucose, the body is forced to burn fat. In these circumstances, fats are converted into fatty acids and ketone bodies that will become the main source of energy for the brain, replacing glucose.

carbohydrates and sugars, has been highlighted. This formula also extends life expectancy in mice, reduces obesity, and appears to help fight cancer.

Another variable that has been manipulated in the laboratory is feeding times: how to distribute mealtimes during the day, or how to modify eating patterns throughout the week. There is enough scientific evidence to show that alternate fasting periods can protect against obesity, cardiovascular disease, diabetes, hypertension, and neurodegeneration. Many versions of what is now called *intermittent fasting* have been tried, in which standard feeding days (when you can eat as much as you want) alternate with fasting days (when you are allowed to drink and eat a maximum of only 20% of your daily calories). In rodents, both alternate-day fasting and eat-stop-eat diet (fasting completely for two days a week) have been reported to extend their life by 30% regardless of weight loss and total caloric intake. Another option, less drastic than not eating for a whole day, would be *time-restricted feeding*. Restricting the hours of eating to approximately 5–8 h throughout the day has equally beneficial effects on health compared to the traditional 3–5 meals spread over 10–12 h. Following certain patterns in eating schedules also seems to have very important beneficial effects on health and aging.

Thus far, we have discussed the knowledge gained from experiments on laboratory animals, especially mice. However, what about humans? What is known? It is too early to draw final conclusions, since there are not enough clinical studies in humans, and this kind of trial has been launched only in the last few years. What does seem to be proven is that people who follow a calorie-restricted diet are protected against obesity and insulin resistance (a condition in which the body's cells do not respond to insulin, and is one of the main features of diabetes), and have levels considered within normal established ranges for different markers of inflammation. Fortunately, we have a study of calorie restriction that was carried out in monkeys, specifically in macaques, and that would represent the most similar situation to humans. This experiment was conducted over a period of more than 20 years in parallel, but independently, in two laboratories in the USA: one at the National Institute on Aging (NIA-Baltimore), and the other at the University of Wisconsin. Initially, these studies showed different conclusions: while the Wisconsin calorie-restricted monkeys lived significantly longer than their peers who had ad libitum access to food, the NIA macaques, even though they ate 30% less, did not have an advantage in terms of their average life expectancy. However, despite not living significantly longer, the incidence of cancer and cardiovascular disease in the NIA macaques was found to be 50% lower in the calorie-restricted monkeys than in those kept ad libitum. One of the keys that may explain these differences in survival lies in the type of food given to the animals. While the monkeys in the Wisconsin study received mostly semiprocessed food with a high percentage of sugars, the Baltimore monkeys enjoyed a more natural diet, rich in nuts and fiber, and low in fat. This difference in food quality is responsible for the fact that the monkeys with ad libitum access to the NIA food, even when eating to satiety, were similar in weight to the macaques kept on a 30% caloric restriction regimen in Wisconsin. Similarly, despite consuming the same number of calories, the Wisconsin food-restricted monkeys were fatter than the NIA-restricted monkeys, demonstrating the importance of food quality on health and longevity in the face of equal caloric intake.

Finally, after several months, both groups agreed to pool and contrast their data and published what are now considered the final results of a joint study on the effect of calorie restriction in monkeys. First and foremost, calorie restriction does extend life expectancy in monkeys. As an example, the researchers describe one of their monkeys, on calorie restriction since the age of 16, an age considered mature for these animals, which has surpassed 43 years of life—an absolute record for this species. That would be equivalent to 130 years for a human being. Additionally, macaques on calorie restriction not only live longer, but also suffer significantly fewer diseases such as cancer, insulin resistance, diabetes, and cardiovascular disease. This comparison also reveals some important nuances, such as limiting calories only extending life when implemented in adult animals, but not from an early or juvenile age. This trial in monkeys is very relevant if we want to implement this type of regimen in humans. It is worth remembering that interventions that work in other animals are not necessarily valid in humans and may even be dangerous.

For some, the promise of living longer by reducing food consumption by 25–50% may be an enticing prospect; for others, it may represent too great a sacrifice. Consequently, the next level in research on aging will be to find compounds that mimic the effects of calorie restriction, for example, imagine a magic pill that has the same benefits without the drawbacks of not eating. We are still a long way off, but there is one compound that shows great potential. *Rapamycin*, a drug known for its immunosuppressive effects and which also reduces insulin-PI3K signaling, is currently considered the most important chemical intervention for extending life expectancy in mice and other mammals. Could it work in humans too? It is too early to say for sure, but it is significant that we are facing a possible revolution of commonly-called *calorie restriction mimetics*[7] such as rapamycin, which could be the key to helping us live longer and with a higher quality of life. Until that happens, by following a healthy diet, avoiding being overweight and staying physically and mentally active, we may increase our chances of extending our life expectancy without having to count every calorie we eat.

---

[7] Rapacimin is the calorie restriction mimetic that has thus far proven to be the most efficient in mice trials. Another compound of great interest to scientists is metformin. This is a drug commonly used for the treatment of type 2 diabetes that increases insulin sensitivity. Although it does not extend life expectancy in many of the organisms in which it has been tested, metformin continues to receive a great deal of attention because it has a very similar effect to calorie restriction on gene expression. Resveratrol, a compound found in red grapes and wine, entered the calorie restriction mimetic scene with a vengeance. At first it seemed that its use was indeed able to delay aging in several laboratory organisms, and reduce the risk of several age-related pathologies. However, little by little it has been seen that the effects of resveratrol are not as wonderful as expected, and currently we have inconsistent publications and very little evidence to support its real benefits on aging, cardiovascular diseases or cancer.

# Bibliography

Baker DJ, Wijshake T, Tchkonia T, Lebrasseur NK, Childs BG, van de Sluis B, Kirkland JL, van Deursen JM (2011) Clearance of p16 Ink4a positive senescent cells delays ageing-associated disorders. Nature 479(7372):232–236

Bernardes de Jesus B, Vera E, Schneeberger K, Tejera AM, Ayuso E, Bosch F, Blasco MA (2012) Telomerase gene therapy in adult and old mice delays aging and increases longevity without increasing cancer. EMBO Mol Med 4(8):691–704

Blackburn EH, Greider CW, Szostak JW (2006) Telomeres and telomerase: the path from maize, Tetrahymena and yeast to human cancer and aging. Nat Med 12:1133–1138

Blasco MA (2007) Telomere length, stem cells and aging. Nat Chem Biol 3(10):640–649

Boonekamp JJ, Simons MJP, Hemerik L, Verhulst S (2013) Telomere length behaves as biomarker of somatic redundancy rather than biological age. Aging Cell 12(2):330–332

Campisi J, D'Adda Di Fagagna F (2007) Cellular senescence: when bad things happen to good cells. Nat Rev Mol Cell Biol 8(9):729–740

Collado M, Blasco MA, Serrano M (2007) Cellular senescence in cancer and aging. Cell 130(2):223–233

Colman RJ, Anderson RM, Johnson SC, Kastman EK, Kosmatka KJ, Beasley TM, Allison DB, Cruzen C, Simmons HA, Kemnitz JW, Weindruch R (2009) Caloric restriction delays disease onset and mortality in rhesus monkeys. Science 325(5937):201–204

De Cabo R, Carmona-Gutierrez D, Bernier M, Hall MN, Madeo F (2014) The search for antiaging interventions: from elixirs to fasting regimens. Cell 157(7):1515–1526

Di Francesco A, Di Germanio C, Bernier M, De Cabo R (2018) A time to fast. Science 362(6416):770–775

Doonan R, McElwee JJ, Matthijssens F, Walker GA, Houthoofd K, Back P, Matscheski A, Vanfleteren JR, Gems D (2008) Against the oxidative damage theory of aging: superoxide dismutases protect against oxidative stress but have little or no effect on life span in Caenorhabditis elegans. Genes Dev 22(23):3236–3241

Fontana L, Partridge L, Longo VD (2010) Extending healthy life span-from yeast to humans. Science 328(5976):321–326

Fontana L, Partridge L (2015) Promoting health and longevity through diet: from model organisms to humans. Cell 161(1):106–118

Gems D, Partridge L (2013) Genetics of longevity in model organisms: debates and paradigm shifts. Annu Rev Physiol 75:621–644

Harman D (1965) The free radical theory of aging: effect of age on serum copper levels. J Gerontol 20:151–153

Harrison DE, Strong R, Sharp ZD, Nelson JF, Astle CM, Flurkey K, Nadon NL, Wilkinson JE, Frenkel K, Carter CS, Pahor M, Javors MA, Fernandez E, Miller RA (2009) Rapamycin fed late in life extends lifespan in genetically heterogeneous mice. Nature 460(7253):392–395

Hopkins BD, Pauli C, Du X, Wang DG, Li X, Wu D, Amadiume SC, Goncalves MD, Hodakoski C, Lundquist MR, Bareja R, Ma Y, Harris EM, Sboner A, Beltran H, Rubin MA, Mukherjee S, Cantley LC (2018) Suppression of insulin feedback enhances the efficacy of PI3K inhibitors. Nature 560(7719):499–503

Kroemer G, López-Otín C, Madeo F, de Cabo R (2018) Carbotoxicity-noxious effects of carbohydrates. Cell 175(3):605–614

Kuban KCK, Levitton A (1994) The effect of vitamin E and beta carotene on the incidence of lung cancer and other cancers in male smokers. New Engl J Med 330(15):1029–1035

Kuningas M, Mooijaart SP, Van Heemst D, Zwaan BJ, Slagboom PE, Westendorp RGJ (2008) Genes encoding longevity: from model organisms to humans. Aging Cell 7(2):270–280

Le Couteur DG, Solon-Biet S, Cogger VC, Mitchell SJ, Senior A, De Cabo R, Raubenheimer D, Simpson SJ (2016) The impact of low-protein high-carbohydrate diets on aging and lifespan. Cell Mol Life Sci 73(6):1237–1252

Lee SH, Min KJ (2013) Caloric restriction and its mimetics. BMB Rep 46(4):181–187

Longo VD, Antebi A, Bartke A, Barzilai N, Brown-Borg HM, Caruso C, Curiel TJ, de Cabo R, Franceschi C, Gems D, Ingram DK, Johnson TE, Kennedy BK, Kenyon C, Klein S, Kopchick JJ, Lepperdinger G, Madeo F, Mirisola MG, Mitchell JR, Passarino G, Rudolph KL, Sedivy JM, Shadel GS, Sinclair DA, Spindler SR, Suh Y, Vijg J, Vinciguerra M, Fontana L (2015) Interventions to slow aging in humans: are we ready? Aging Cell 14(4):497–510

López-Otín C, Blasco MA, Partridge L, Serrano M, Kroemer G (2013) The hallmarks of aging. Cell 153(6):1194–1217

Mackenbach JP, Looman CW (2013) Life expectancy and national income in Europe, 1900–2008: an update of Preston's analysis. Int J Epidemiol 42(4):1100–1110

Mattson MP, Allison DB, Fontana L, Harvie M, Longo VD, Malaisse WJ, Mosley M, Notterpek L, Ravussin E, Scheer FA, Seyfried TN, Varady KA, Panda S (2014) Meal frequency and timing in health and disease. Proc Natl Acad Sci 2111(47):16647–16653

Mattison JA, Colman RJ, Beasley TM, Allison DB, Kemnitz JW, Roth GS, Ingram DK, Weindruch R, de Cabo R, Anderson RM (2017) Caloric restriction improves health and survival of rhesus monkeys. Nat Commun 8:1–12

Mattison JA, Roth GS, Mark Beasley T, Tilmont EM, Handy AM, Herbert RL, Longo DL, Allison DB, Young JE, Bryant M, Barnard D, Ward WF, Qi W, Ingram DK, de Cabo R (2012) Impact of caloric restriction on health and survival in rhesus monkeys from the NIA study. Nature 489(7415):318–321

McCay CM, Crowell MF (1934) Prolonging the life span. Sci Mon 39(5):405–414

Miller RA, Buehner G, Chang Y, Harper JM, Sigler R, Smith-Wheelock M (2005) Methionine-deficient diet extends mouse lifespan, slows immune and lens aging, alters glucose, T4, IGF-I and insulin levels, and increases hepatocyte MIF levels and stress resistance. Aging Cell 4(3):119–125

Mitchell SJ, Bernier M, Mattison JA, Aon MA, Kaiser TA, Anson RM, Ikeno Y, Anderson RM, Ingram DK, de Cabo R (2008) Daily fasting improves health and survival in male mice independent of diet composition and calories. Cell Metab, 1–8

Mitchell SJ, Madrigal Matute J, Scheibye-Knudsen M, Fang E, Aon M, González-Reyes JA, Cortassa S, Kaushik S, Gonzalez-Freire M, Patel B, Wahl D, Ali A, Calvo-Rubio M, Burón MI, Guiterrez V, Ward TM, Palacios HH, Cai H, Frederick DW, Hine C, Broeskamp F, Habering L, Dawson J, Beasley TM, Wan J, Ikeno Y, Hubbard G, Becker KG, Zhang Y, Bohr VA, Longo DL, Navas P, Ferrucci L, Sinclair DA, Cohen P, Egan JM, Mitchell JR, Baur JA, Allison DB, Anson RM, Villalba JM, Madeo F, Cuervo AM, Pearson KJ, Ingram DK, Bernier M, de Cabo R (2016) Effects of sex, strain, and energy intake on hallmarks of aging in mice. Cell Metab 23(6):1093–1112

Most J, Tosti V, Redman LM, Fontana L (2017) Calorie restriction in humans: an update. Ageing Res Rev 39:36–45

Omenn GS, Goodman GE, Thornquist MD, Balmes J, Cullen MR, Glass A, Keogh JP, Meyskens FL, Valanis B, Williams JH, Barnhart S, Hammar S (1996) Effects of a combination of beta carotene and vitamin A on lung cancer and cardiovascular disease. N Engl J Med 334(18):1150–1155

Perez VI, Van Remmen H, Bokov A, Epstein CJ, Vijg J, Richardson A (2009) The overexpression of major antioxidant enzymes does not extend the lifespan of mice. Aging Cell 8(1):73–75

Perls TT, Wilmoth J, Levenson R, Drinkwater M, Cohen M, Bogan H, Joyce E, Brewster S, Kunkel L, Puca A (2002) Life-long sustained mortality advantage of siblings of centenarians. Proc Natl Acad Sci 99(12):8442–8447

Ristow M, Schmeisser S (2011) Extending life span by increasing oxidative stress. Free Radic Biol Med 51(2):327–336

Roberts MN, Wallace M, Tomilov AA, Zhou Z, Marcotte GR, Tran D, Perez G, Gutierrez-Casado E, Koike S, Knotts TA, Imai DM, Griffey SM, Kim K, Hagopian K, McMackin MZ, Haj FG, Baar K, Cortopassi GA, Ramsey JJ, Lopez-Dominguez JA (2017) A ketogenic diet extends longevity and healthspan in adult mice. Cell Metab 26(3):539-546.e5

Samaras TT, Elrick H (2002) Height, body size, and longevity: is smaller better for the human body? West J Med 176(3):206–208

Sebastiani P, Solovieff N, DeWan AT, Walsh KM, Puca A, Hartley SW, Melista E, Andersen S, Dworkis DA, Wilk JB, Myers RH, Steinberg MH, Montano M, Baldwin CT, Hoh J, Perls TT (2012) Genetic signatures of exceptional longevity in humans. PLoS ONE 7(1):e29848

Simpson SJ, Le Couteur DG, Raubenheimer D, Solon-Biet SM, Cooney GJ, Cogger VC, Fontana L (2017) Dietary protein, aging and nutritional geometry. Ageing Res Rev 39:78–86

Solon-Biet SM, Mitchell SJ, De CR, Raubenheimer D, Le CDG, Simpson SJ (2015) Macronutrients and caloric intake in health and longevity. J Endocrinol 226(1):R17-28

Van Remmen H, Ikeno Y, Hamilton M, Pahlavani M, Wolf N, Thorpe SR, Alderson NL, Baynes JW, Epstein CJ, Huang TT, Nelson J, Strong R, Richardson A (2003) Life-long reduction in MnSOD activity results in increased DNA damage and higher incidence of cancer but does not accelerate aging. Physiol Genomics 16(1):29–37

Westendorp RGJ, Van Heemst D, Rozing MP, Frölich M, Mooijaart SP, Blauw GJ, Beekman M, Heijmans BT, de Craen AJ, Slagboom PE; Leiden Longevity Study Group (2009) Nonagenarian siblings and their offspring display lower risk of mortality and morbidity than sporadic nonagenarians: The Leiden longevity study. J Am Geriatr Soc 57(9):1634–1637

Zimmerman JA, Malloy V, Krajcik R, Orentreich N (2003) Nutritional control of aging. Exp Gerontol 38(1–2):47–52

**Elena López Guadamillas** holds a degree in Biology, a master's degree in Molecular Biomedicine and a Ph.D. in Biological Sciences from Universidad Autónoma de Madrid (UAM). Following her doctoral research at the Spanish National Cancer Research Centre (CNIO) on cancer and obesity, she undertook her postdoctoral studies at the University College of London (UCL) Cancer Institute. Where she investigated the role of PI3Kd in the tumor microenvironment. She is author of more than a dozen publications in national and international scientific conferences and journals.

# Food Behavior: A Breadcrumb Trail to Addiction

Carmen Rodríguez Rivera ⓘ

Where can a trail of breadcrumbs lead? Can you tell from the first crumb, what will be the final destination of the journey?

Obesity is an increasingly prevalent pathology that constitutes a real public health problem. Leaving COVID-19 aside, it has come to be described as the "Pandemic of the twenty-first century" by the World Health Organization (WHO), with more than a third of the world's population suffering from it; when we are talking about the child population, it is one in ten. Obesity leads to a high number of comorbidities, as well as a significant shortening of life expectancy, and was the main cause of approximately 3.4 million deaths as of 2010.

Although somewhat controversial, the main criterion for diagnosing obesity is currently based on body mass index (BMI), which is calculated as body weight in kilograms divided by height in meters squared: $\text{BMI} = \frac{\text{weight (kg)}}{\text{height (m)}^2}$. The WHO establishes different ranges to classify people according to their BMI (Table 1).

C. Rodríguez Rivera (✉)
Area of Pharmacology, Nutrition and Bromatology, Department of Basic Health Sciences, Unidad Asociada de I+D+i al Instituto de Química Médica (IQM-CSIC), Universidad Rey Juan Carlos, Madrid, Spain
e-mail: carmen.rodriguez@urjc.es

High Performance Experimental Pharmacology Research Group (PHARMAKOM), Universidad Rey Juan Carlos, Alcorcón, Spain

Grupo Multidisciplinar de Investigación y Tratamiento del Dolor (i+DOL), Alcorcón, Spain

© The Author(s), under exclusive license to Springer Nature Switzerland AG 2024
M. M. Garcia (ed.), *Tales of Discovery*, https://doi.org/10.1007/978-3-031-47620-4_7

**Table 1** World Health
Organization classification
according to body mass index
(BMI)

| Classification | BMI |
| --- | --- |
| Low weight | < 18.5 |
| Normal weight | 18.5–24.9 |
| Overweight | 25.0–29.9 |
| Type I obesity | 30.0–34.9 |
| Type II obesity | 35.0–39.9 |
| Type III obesity (morbid obesity) | ≥ 40.0 |

## Multiple Crumbs Make Their Way

Obese individuals share a number of risks, regardless of the cause of their disease. Obesity produces a high accumulation of fat in the body; this is harmful because a state of chronic inflammation is maintained over time. Adipose tissue has the capacity to increase the amount of triglycerides it stores. When the quantity of fatty acids exceeds this capacity, a phenomenon known as *lipotoxicity* occurs; triglycerides that cannot be stored are distributed to the rest of the tissues through the circulatory system, causing accumulations of fat in organs where they should not be found, and eventually leading to other pathologies. Lipotoxicity can affect organs such as the liver, ultimately leading to a fatty liver that loses its ability to metabolize drugs, toxins and other substances. It can also affect arteries and veins, where the accumulation of fat decreases the ability of the vessels to allow blood to pass through them, ultimately causing hypertension. Over time, it can block the passage of blood in such a way as to cause ischemia, heart attacks or even death.

That said, adipose tissue has endocrine and metabolic functions that are modified by the disease. While the small adipocytes of a normal-weight individual would promote correct metabolic functioning (correct energy functioning), the thickened adipocytes of obese individuals would be able to stimulate the inflammatory process and the release of adipokines, such as IL-6 or TNFα. This, in turn, would contribute to the progression of insulin resistance and other types of alterations that affect various organs and systems, including the central nervous system.

Apart from these circumstances common to all patients with obesity, the different causes that can give rise to the disease must be considered in order to design personalized therapeutic approaches. Today we know that several factors contribute jointly, but unequally, to the genesis of the problem: genetic factors; and, nutritional and behavioral conditioning factors, i.e., ingestive behaviors and sedentary lifestyle—or, our trail of breadcrumbs.

From a nutritional point of view, the responsibility for weight gain in obese individuals is often attributed to net caloric intake; these patients exhibit a positive energy balance between what is consumed and what is expended. However, the nature of what is ingested can also play an extremely important role. In the last three decades, many societies have undergone substantial changes in their lifestyles and diets, shifting from a healthy and balanced diet to an exacerbated consumption of foods with a high sugar content and processed fats, which can generate changes in brain physiology. These changes are related to the behavioral factors involved in obesity, since the consumption of certain foods could be the cause of alterations- in the central nervous system that favor impulsivity and hyperphagia, thus perpetuating the problem—similar to what happens in drug addiction.

Increased impulsivity in decision-making seems to be present in an increasing proportion of the obese population, to the extent that much discussion as to whether it is the breadcrumbs that lead to obesity or whether obesity itself is just another crumb. Even if diet is the common causality to both, is obesity the cause of impulsivity, or vice versa? Recent studies in animals and humans have linked high-fat diets with a greater development of impulsive behaviors. Authors such as Steele and colleagues in 2017 found that rats fed a high-fat diet showed a tendency to choose shorter but faster rewards over longer and slower rewards, a result that demonstrates an increase in impulsivity in these animals compared to those fed a standard control diet. Authors Lumley and colleagues, analyzed impulsivity in humans in 2016 using questionnaires and scales of intake of certain foods. They observed a significant correlation between impulsivity and the consumption of foods and beverages rich in sugars and processed fats.

It is generally considered that in impulsive/compulsive disorders, impulsivity is an initial risk factor and, as described by authors Alguacil and Gonzalez-Martín, compulsivity is a later state that is due to a lack of impulse control. This triggers pathological behaviors in relation to certain stimuli, such as those related to the activity of eating. This is a very recognizable behavioral pattern in the case of substance use disorders such as drugs of abuse, where it is easy to observe that the search for and consumption of a drug becomes increasingly important for the consumer, who ends up developing an addiction. In recent years, the parallel between this type of addiction and the pattern of consumption of certain foods, both in animals and humans, has been highlighted, leading to the formulation of the concept of *food addiction* and the postulation that a proportion of obese individuals may be affected by this problem.

Similar to the above-mentioned impulsivity, the shift to a high-calorie, appetizing diet could be one of the triggers of this type of addictive behavior towards food, as authors Alsiö and collaborators have discovered. However, where and how far does this path of crumbs lead us?

## How Did We Get to Addiction?

Ingestion is a very complex behavior that results from an interaction between metabolic signals, and cognitive and emotional mechanisms strongly influenced by the environment. For proper energy homeostasis to occur, hormones such as insulin, ghrelin, and other peptides (e.g., NPY, AgRP, POMC and CART) are released by the gut, stomach, pancreas, or adipose tissue, controlling satiety or hunger signals at the hypothalamic level.

Basically, in situations of energy deficit, ghrelin is secreted from the stomach, leading to the stimulation of receptors in the neurons of the *hypothalamic arcuate nucleus*, and this in turn increases the release of NPY and AgRP, neuropeptides, which promote intake. In contrast, in situations of energetic excess, leptin and insulin levels increase; these hormones inhibit the release of NPY and AgRP, and promote the release of satiating peptides POMC and CART, which inhibit ingestion. It could thus be said that the homeostatic need to ingest food would be regulated by this prominent hypothalamic pathway.

However, the desire to eat palatable foods is regulated by the hedonic system, which can incentivize intake regardless of energy status. The hedonic system comprises the brain reward circuit, or mesolimbic dopaminergic circuit (Fig. 1), in which the palatable taste of certain foods triggers the release of dopamine, producing pleasurable sensations in exactly the same way as drugs of abuse do.[1]

In physiological situations, both homeostatic and hedonic systems are closely integrated and interregulated (Fig. 2). Thus, for example, an intake initiated by stimulus of the hedonic pathway, that is, by the mere pleasure that eating produces, may be limited by the influence of the homeostatic system, which represses the hedonic system when one's meal is too copious. However, these interregulatory mechanisms could be affected in pathological situations, so an inappropriate regulation of behavior would lead to abnormal increases or decreases in body weight.

Thus, in aberrant situations, such as those produced by various "breadcrumbs," a response to food can be triggered that would consist of an addictive type of disorder, accompanied by a lack of control of the cerebral reinforcement/reward system, in other words, an altered cognitive processing of the contextual cues associated with food. On the one hand, the homeostatic system, which tells us when to eat (we should not continue eating food when our energy needs have been satisfied), would lose its ability to inhibit the hedonic system (the pleasure system). On the other hand, in the hedonic system changes triggered by excessive stimulation, and maintained over time, would lead to a deficient function of: the reinforcement/reward system; origin of impulsive/compulsive behaviors that appear in various psychiatric diseases such as addictions; and other types of diseases such as Tourette's syndrome or Attention Deficit Hyperactivity Disorder (ADHD). The various symptoms we would find are manifestations of *reward deficiency syndrome*.

These symptoms include physical and psychological dependence on food. The first would be what is traditionally known as *withdrawal syndrome*. The organism itself would be requesting to consume the *drug*, since it has already become

---

[1] Stimulation of the hedonic system: When a reinforcer stimulates the mesolimbic system, dopamine release occurs from the ventral tegmental area (VTA), towards the Nucleus Accumbens (NAc), from this nucleus the signal propagates towards the prefrontal cortex (CPF), responsible on the one hand for the cognitive control of the activity of the VTA and NAc neurons through the release of GABA and glutamate, and on the other hand for the communication with the orbitofrontal cortex (COF), which ultimately determines the intake.

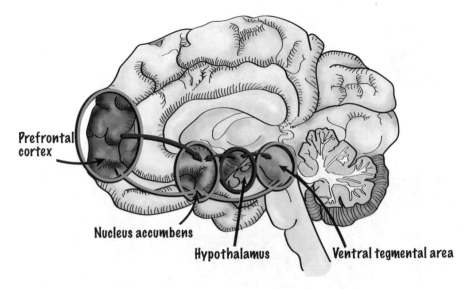

**Fig. 1** **Diagram of the brain circuit of reward and dopaminergic signaling.** Blue represents the areas of the brain reward circuit: Ventral Tegmental Area, Nucleus Accumbens (reward and addiction) and Prefrontal Cortex (decision making and executive functions), as well as the projections of such circuit. Red represents the hypothalamus, which integrates homeostatic and hedonic information, and is able to inhibit such circuits in physiological conditions with its extensions to the ventral tegmental area and nucleus accumbens; inhibition can also be performed by the prefrontal cortex

accustomed to its consumption; in its absence, there are various physical responses such as hypoglycemia, excessive sweating, or even tremors. The second refers to *cravings*—the irrepressible desire to consume a substance. This would explain the numerous behavioral parallels that have been found between the phenomena of food and drug addiction, both in humans and in experimental models. Tasty foods would act like drugs of abuse, causing an increase in the release of dopamine in certain areas of the brain, such as the nucleus accumbens. This in turn would generate a pleasurable sensation. When access to these diets rich in sugars and fats is prolonged over time, neurochemical changes become chronic and alter the responsiveness of the brain's reward circuit, leading to excessive stimulation and an addictive disorder.

The key question then is this: Which of us would reach the end of that breadcrumb trail? It is estimated that food addiction occurs in one in five obese patients, and if we are talking about morbidly obese patients (IBM $\geq$ 40), the incidence could be as high as 54%.

Additionally, it has been suggested that patients undergoing bariatric surgery to treat obesity may be at increased risk of experiencing the brain alterations that lead to persistent addictive behavior. In this case, because the patient is not able to eat the amount of food at which he/she normally achieved the desired level of pleasure, there could be a phenomenon of *addiction transfer*, in which the previous

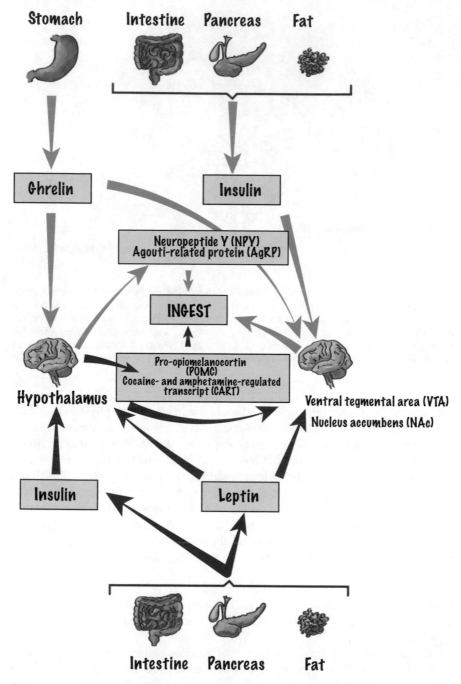

**Fig. 2 Integration of hedonic and homeostatic systems for the regulation of intake**. The green lines represent the excitatory and maroon lines the inhibitory pathways of the homeostatic (hypothalamus mediating the release of POMC and CART, and NPY and AgRP) and hedonic (VTA- Ventral Tegmental Area; and NAC- Nucleus Accumbens) systems for the control of intake

addiction to food would be transformed into an addiction to drugs of abuse (mainly alcohol) or even to gambling. In this sense, it is estimated that the proportion of new drug users among patients undergoing bariatric surgery could reach almost 90%.

In view of the above, it is evident that there is a need to progress in the identification and biological characterization of the addictive phenotype of obesity. In recent years, several authors have used the Yale Food Addiction Scale (YFAS) to diagnose food addiction in the same way as other types of addictions are diagnosed. However, no biological markers have yet been indubitably identified to associate with a possible addictive phenotype of obesity.

That first "crumb" is often not dropped by us, but by our ancestors. In some people, there is a genetic predisposition such that, in the presence of more crumbs, they are more vulnerable to developing addictions. The second crumb could, in some cases, have been dropped by our mother during gestation. At this stage, the brain is developing, and a small disturbance is capable of causing changes that are sometimes irreversible. Undernutrition (the lack of food) would be one such disturbance, as it can trigger numerous mental pathologies, including schizophrenia, attention deficit hyperactivity disorder, or addictions such as food addiction. Undernutrition during this stage would be capable of causing an increase in the amount of food we eat, as well as increasing the ability of food to generate pleasurable sensations during adulthood. It seems that an increased synthesis of dopamine after food intake caused by an aberrant maturation of the neurons of the hedonic system would be behind this. Following this same line, it has been observed that among the descendants of pregnant mothers who suffered the *Dutch famine* at the end of World War II, there is a higher prevalence people addicted to drugs of abuse. The origin of the third crumb is more questionable, as it may or may not depend on oneself. Here we are talking about the economic and social circumstances that surround us and that have a determining influence on our own decisions. Finally, in the case of food addiction, the fourth key crumb would represent those foods rich in fats eaten on an occasional basis, capable of hyperstimulating the hedonic system, and causing much more pleasurable sensations than normal foods do. Although transitory, this would stimulate an excessive motivation for the consumption of this type of food. However, it would only be chronic consumption that would cause more persistent changes in the brain, such as a decrease in the stimulation produced in areas of the pleasure system.

Without even realizing it, our "trail of breadcrumbs" can bring us closer and closer to a very real danger: addiction.

## Bibliography

Alguacil LF, González-Martín C (2015) Target identification and validation in brain reward dysfunction. Drug Discovery Today 20(3):347–352

Blum K, Febo M, Badgaiyan RD, Demetrovics Z, Simpatico T, Fahlke C, M O-B, Li M, Dushaj K, Gold MS (2017) Common neurogenetic diagnosis and meso-limbic manipulation of hypodopaminergic function in reward deficiency syndrome (RDS): changing the recovery landscape. Current Neuropharmacol 15(1):184–94

da Silva AAM, Borba TKF, de Almeida Lira L, Cavalcante TCF, de Freitas MFL, Leandro CG, do Nascimento E, De Souza SL (2013) Perinatal undernutrition stimulates seeking food reward. Int J Dev Neurosci 31(5):334–41

da Silva AAM, Oliveira MM, Cavalcante TCF, do Amaral Almeida LC, de Souza JA, da Silva M C, de Souza SL (2016) Low protein diet during gestation and lactation increases food reward seeking but does not modify sucrose taste reactivity in adult female rats. Int J Dev Neurosci 49:50–9

de Melo Martimiano PH, da Silva GR, Coimbra VFSA, Matos RJB, de Souza BFP, da Silva AAM, de Melo DDCB, de Souza SL, de Freitas MFL (2015) Perinatal malnutrition stimulates motivation through reward and enhances drd1a receptor expression in the ventral striatum of adult mice. Pharmacol Biochem Behav 134:106–114

Figlewicz DP, Jay JL, Acheson MA, Magrisso IJ, West CH, Zavosh A, Benoit SC, Davis JF (2013) Moderate high fat diet increases sucrose self-administration in young rats. Appetite 61(1):19–29

Gordon EL, Ariel-Donges AH, Bauman V, Merlo LJ (2018) What is the evidence for "food addiction?" A systematic review. Nutrients 10(4):477

Haslam DW, James WPT (2005) Obesity. The Lancet 366(9492):1197–1209

Hruby A, Hu FB (2015) The epidemiology of obesity: a big picture. Pharmacoeconomics 33(7):673–689

la Fleur SE, van Rozen AJ, Luijendijk MCM, Groeneweg F, Adan RA (2010) A free-choice high-fat high-sugar diet induces changes in arcuate neuropeptide expression that support hyperphagia. Int J Obes 34(3):537–46. Available at: https://doi.org/10.1038/ijo.2009.257

Nathan PJ, Bullmore ET (2009) From taste hedonics to motivational drive: central μ-opioid receptors and binge-eating behaviour. Int J Neuropsychopharmacol. 12(7):995–1008. Available at: https://doi.org/10.1017/S146114570900039X.

Nestler EJ (2004) Molecular mechanisms of drug addiction. Neuropharmacology 47(1):24–32

Nestler EJ (2005) The neurobiology of cocaine addiction. Sci Pract Perspect 3(1):4–10. Available at: http://www.ncbi.nlm.nih.gov/pmc/articles/PMC2851032/

Ng M, Fleming T, Robinson M, Thomson B, Graetz N, Margono C, Mullany EC, Biryukov S, Abbafati C, Abera SF, Abraham JP, Abu-Rmeileh NM, Achoki T, AlBuhairan FS, Alemu ZA, Alfonso R, Ali MK, Ali R, Guzman NA, Ammar W, Anwari P, Banerjee A, Barquera GH et al (2014) Global, regional, and national prevalence of overweight and obesity in children and adults during 1980–2013: a systematic analysis for the Global Burden of Disease Study 2013. The Lancet 384(9945):766–781

Opolski M, Chur-Hansen A, Wittert G (2015) The eating-related behaviours, disorders and expectations of candidates for bariatric surgery. Clinical Obesity. 5(4):165–197

Palmiter RD (2007) Is dopamine a physiologically relevant mediator of feeding behavior? Trends Neurosci 30(8):375–381

Steffen K, Engle SG, Wonderlich J, Pollert G, Sondag C (2015) Alcohol and other addictive disorders following bariatric surgery: prevalence, risk factors and possible etiologies. Eur Eat Disord Rev 23(6):442–450

Stevens GA, Singh GM, Lu Y, Danaei G, Lin JK, Finucane MM, Bahalim AN, McIntire RK, Gutierrez HR, Cowan M, Paciorek CJ, Farzadfar F, Riley L, Ezzati M (2012) Global burden of metabolic risk factors for chronic diseases collaborating group (Body Mass Index). National, regional, and global trends in adult overweight and obesity prevalences. Population Health Metrics. 10(1):22

Virtue S, Vidal-Puig A (2008) It is not how fat you are, it is what you do with it that counts. PLOS Biol 6(9):e237. Available at: https://doi.org/10.1371/journal.pbio.0060237

Volkow ND, O'Brien CP (2007) Issues for DSM-V: should obesity be included as a brain disorder? Am J Psychiatry 164(5):708–710

Volkow N, Wang G-J, Tomasi D, Baler R (2013) The addictive dimensionality of obesity. Biol Psychiat 73(9):811–818

Weisberg SP, McCann D, Desai M, Rosenbaum M, Leibel RL, Ferrante AW Jr (2003) Obesity is associated with macrophage accumulation in adipose tissue. J Clin Investig 112(12):1796–1808

**Carmen Rodríguez Rivera** is a graduate, master and doctor in Pharmacy from Universidad San Pablo CEU and after participating in the university teaching in the areas of Pharmacology and Toxicology of the Department of Pharmaceutical and Health Sciences of same University, she is currently an Assistant Professor and Researcher in the Area of Pharmacology, Nutrition and Bromatology of Universidad Rey Juan Carlos. She has conducted different internships at national and international research centers and is the author of several publications in books, journals and scientific conferences both nationally and internationally. Her research work has focused on the determination of biomarkers and the study of the physiology of addictions. She belongs to the High Performance Research Group in Experimental Pharmacology (PHARMAKOM) of Universidad Rey Juan Carlos, and is a member of the Spanish Pain Society (SED). As a disseminator, she has experience as a speaker in different activities such as *Science Week* and the *European Researchers' Night*.

# Global Opioid Crisis: Two Sides of the Same Coin

Miguel M. Garcia⬤, Nancy Antonieta Paniagua Lora⬤, and Eva Mercado Delgado⬤

One of the most influential women in the history of the Conquest of America was Malinalli, also known as *La Malinche* or *Marina*, a Nahuatl princess who was betrayed by her family and sold as a slave to the governor of Pontochán, a Mayan city located in the current state of Tabasco. After the Spaniards defeated Cortes, Malinalli's luck began to change. She was able to speak Nahuatl, Mayan, Mexica and Spanish, and knew their respective customs; she became the interpreter between Cortés and the Spaniards, and the different peoples of ancient Mexico, which was valuable in the defeat and conquest of the Aztecs. Consequently, a term took root in Mexican lore known as *malinchismo*, used to refer to those people who

M. M. Garcia (✉) · N. A. Paniagua Lora · E. Mercado Delgado
Area of Pharmacology, Nutrition and Bromatology, Department of Basic Health Sciences,
Universidad Rey Juan Carlos, Unidad Asociada I+D+I Instituto de Química Médica (IQM)
CSIC-URJC, Madrid, Spain
e-mail: miguelangel.garcia@urjc.es

N. A. Paniagua Lora
e-mail: nancy.paniagua@urjc.es

E. Mercado Delgado
e-mail: emercado.pex@sanitas.es

M. M. Garcia · N. A. Paniagua Lora
High Performance Experimental Pharmacology Research Group (PHARMAKOM), Universidad
Rey Juan Carlos, Alcorcón, Spain

M. M. Garcia · N. A. Paniagua Lora · E. Mercado Delgado
Grupo Multidisciplinar de Investigación y Tratamiento del Dolor (i+DOL), Alcorcón, Spain

E. Mercado Delgado
Pain Management Unit, Hospital Universitario Sanitas Virgen del Mar, Madrid, Spain

reject their own and embrace the foreign. Whether it is a pluralistic sentiment or something linked to the Hispanic world, the truth is that many of us often tend to think that the "the grass is always greener on the other side". However, is it truly so?

In 1680, the English physician Thomas Sydenham said: "Of all the remedies that God has given to man to alleviate his suffering, none is so universal and efficacious as opium. If we were to throw all medicines into the sea, except opium, it would be a great misfortune to fish and a great benefit to mankind." Years later, in 1700, the Welsh physician John Jones pronounced in similar terms, "Opium often takes away pain by distraction and relaxation caused by pleasure and its incompatibility with pain; it prevents and removes heaviness, fear, anxieties, ill-temper and uneasiness". What are opioid analgesics, and where do they come from? From a pharmacological point of view, *opiate* refers to products derived from opium, perhaps better known as opium poppy, royal poppy, or white poppy.[1] The term *opioid* is left for those endogenous or synthetic products created from modifications of the structure of morphine, the main analgesic compound of the opium poppy. However, both terms are often used interchangeably for synthetic, endogenous, and opium-derived substances.

But wait. *Endogenous* products? Does this mean that we have de facto opioids in our body? Let us start from the assumption that a pipe smoker who is inhaling opium notices a number of effects. This is because compounds from opium bind more or less selectively in the body, specifically to receptors or molecular targets that have been commonly referred to as "opioid receptors." However, although opium has been used for more than five thousand years[2] by the traditional medicine of many civilizations, it was not until 1975 that science discovered that the human body has a whole system of opioid receptors and its own molecules (endogenous opioids) with actions similar to those of the opium poppy, as well as a set of enzymes necessary for its production, transformation, and metabolization (elimination). Endogenous opioids, however, are not useful as drugs since they are peptides that are quickly and easily metabolized. This is why natural or synthetic nonpeptide opioids of an amino acid nature are used.[3]

---

[1] Although all Papaveraceae contain opiates, only the white poppy, *Papaver somniferum*, contains sufficient quantities to be isolated in relatively significant concentrations.

[2] Although the first written reference to the use of opiates for the treatment of pain is found in Babylonian culture, on clay tablets from Nippur, ~2250 B.C., opium use appears to have been prevalent throughout the Mediterranean rim; still-active poppy remnants over three thousand years old have been found in Egyptian tombs.

[3] Proteins are made up of peptides, and these in turn are made up of amino acids. Due to their larger size and the fact that they can be degraded into simpler molecules, peptides are more susceptible to interact with and be attacked (degraded) by molecules of the organism.

## Different Opioids, Different Pharmacology

It has long been thought that the best way to increase the efficacy of a drug is to optimize the dose and frequency of administration; however, this has the consequence of being able to induce greater adverse effects as well. In fact, opioid receptors and ligands are expressed not only throughout the nervous system (peripheral and central), but also in other non-nerve tissues, such as joints, muscle, immune and glial, lung, kidney, and digestive tract. Therefore, both beneficial and toxic effects (due to excess drug in the body) can affect a multitude of organs, e.g., in addition to pain relief, they produce constipation.

How does opioid analgesia work? Let us start from the premise that, in every cell in resting conditions, there is a potential difference (voltage) between both sides of the cell surface: the inner side in direct contact with the cell membrane is mainly occupied by negative charges; the outer side is mainly occupied by positive charges. In a nerve cell, the binding of opioid molecules to receptors on their membranes causes a conformational change in them that allows interaction with other actors inside the cell. This signal transduction culminates in the closure of calcium channels ($Ca^{2+}$) on the one side, and the opening of potassium channels ($K^+$) on the other. Calcium is usually stored in organelles within the cell, and plays an essential role in the fusion of vesicles containing neurotransmitters to the cell membrane, being released into the intersynaptic space. Thus, as long as the calcium channels are closed, there will be little calcium in the cytosol. On the other hand, the intracellular potassium concentration is almost thirty times higher than the extracellular concentration (140 mM versus 5 mM). This means that when potassium channels open, a large quantity of positive charges leave the cell, contributing to a higher electronegativity on the inner side of the cell membrane. This is why, in the presence of an opioid, when a nerve impulse arrives, the change in polarity on both sides of the membrane (depolarization) is less likely to occur. This is because the intracellular side has become more electronegative than in basal conditions, and although the entry of positive ions into the cellular interior ($Na^+$) occurs, the closure of calcium channels and opening of potassium channels prevent a voltage reversal between both sides of the cell membrane. Therefore, opioids would be able to decrease the release of neurotransmitters from the presynaptic neuron, as well as the excitability of the postsynaptic neuron (Fig. 1).

All opioid drugs apparently share this common mechanism of action. However, they do not all have the same pharmacokinetics; they do not all undergo the same actions by the body, or even the same pharmacodynamics; they do not all have the same effect on the body. Slight differences in their chemical structure mean that different opioids give rise to different metabolites, bioavailability, distribution, affinity, selectivity and, ultimately, different effectiveness. In addition, the existence of different types and subtypes of receptors (MOR, DOR, KOR, and ORL1), whose expression varies throughout the different tissues and under certain conditions of the cellular environment, contributes to morphine, oxycodone, fentanyl, or tramadol having a different analgesic efficacy for different types of pain or pathophysiological conditions. The truth is that, in many cases, there is still no

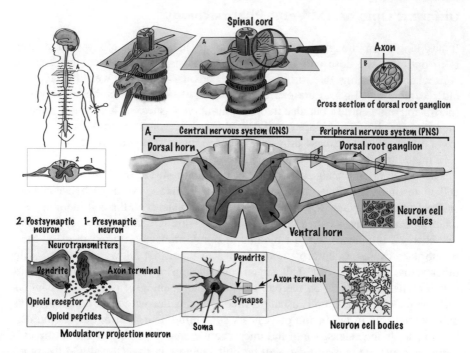

**Fig. 1** **Mechanism of opioid action in the dorsal horn of the spinal cord. A. Cross section of the spinal cord. B. Cross section of the dorsal branch of a peripheral nerve**. Opioid peptides are produced in the neuronal bodies (somas), located in the ganglia of the spinal roots, and mature along the axon in both directions: towards the central nervous system (CNS), and towards the periphery (PNS). They are responsible for modulating the nociceptive signal. In the dorsal horn of the spinal cord, opioid peptides can be released by the different neurons involved in the synapse. The binding of these endogenous opioids to the cell surface receptors of the presynaptic (1) or postsynaptic (2) neuron will decrease neurotransmitter release or neuronal excitability, respectively

clear criteriaon when to administer one or another opioid, and the physician's own clinical experience is limiting factor in the choice: medicine based on experience.

The distribution and passage of individual drug molecules through biological membranes has been especially studied in recent years. Cell membranes are mainly composed of lipids, so molecules that are more lipid-soluble will be able to fuse with them, and pass through them more easily. With regard to its composition and function, the blood–brain barrier consists of a lining formed by several layers of overlapping cells (endothelial, epithelial and glial) that regulate and limit the transit of substances to and from the central nervous system. This is done passively by offering physical resistance to the transit of substances. Only those molecules that are small enough to pass through the junctions between cells (i.e., are very lipid-soluble) will be able to access the central nervous system. However, this flow of substances is also actively regulated. Membrane proteins facilitate or hinder the entry of substances by serving as specific transporters, or by selectively retaining and expelling nutrients, electrolytes, neurotransmitters, toxins, and drugs.

These proteins are the uptake and efflux transporters, and membrane glycoproteins. Animal studies have shown that substances that can interact with P-glycoprotein (P-gp), located on the luminal side of the endothelial cells of the blood–brain barrier (the one that faces the inside of the vessel), are retained and therefore are hindered in their passage. While many opioids (morphine, methadone, loperamide, etc.) are substrates of P-glycoprotein; oxycodone is not. The result is that even though morphine and oxycodone have similar lipid solubility, the concentration of morphine in the brain is lower than its concentration in plasma; in contrast, oxycodone can reach very high concentrations in the central nervous system.[4]

Therefore, the transport of opioid analgesics across the blood–brain barrier largely, but not exclusively, determines their therapeutic effects. Other additional mechanisms must be at work because oxycodone has a lower affinity than morphine for the MOR opioid receptor. When both are administered directly into the central nervous system (intracerebroventricular administration), that is, when the blood–brain barrier with its associated proteins and secondary metabolites no longer intervene, oxycodone has a greater analgesia than morphine. One possible explanation is that a lower affinity, or low binding strength to its receptor would allow oxycodone to be immediately released to act on another receptor. For the moment, however, that is only a theory.

It is also known that there are different vehicles for drugs involved in transport into and out of not only the blood–brain barrier but also the liver and kidney. This would explain the higher metabolic and elimination rates for some drugs than for others. Furthermore, even though they are all opioids, sharing a typical tricyclic structure, different opioids give rise to different proportions of active and inactive metabolites. For example, the metabolization of morphine leads to morphine-6-glucuronide (M6G, active metabolite) and morphine-3-glucuronide (M3G, inactive metabolite). The latter in particular seems to be paradoxically involved in opioid-induced hyperalgesia, that is, the administration of morphine would cause analgesia, but at the same time, if the concentration of M3G were high enough, when metabolized it could induce hyperalgesia and pain. This would explain why the phenomenon of opioid-induced hyperalgesia occurs with morphine or codeine, but does not occur with other drugs (Fig. 2).

The result of all of the above is a different efficacy, potency, and presentation of adverse effects, depending on the drug administered. In addition, the activation of MOR receptors in areas of the brain related to pain-signaling has been shown to be modified differently depending on the opioid used. Theoretically, different opioids can present different profiles in different types of pain. All conditions and circumstances that share the common symptom of pain do not necessarily share the same pathophysiological mechanisms.

Even today, the information available for prescribing opioid drugs is controversial, and can lead to inappropriate use or, conversely, unwarranted fear.

---

[4] Recent studies have discovered a pyrilamine transporter that allows a facilitated transport of oxycodone across the blood–brain barrier, and plays an added role in its therapeutic efficacy.

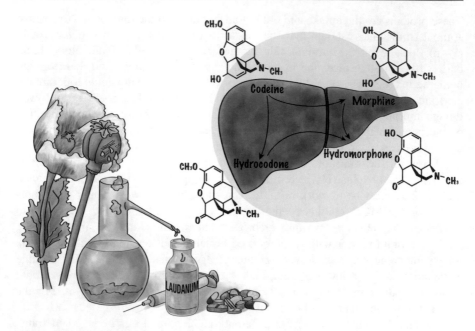

**Fig. 2  Opioid metabolism.** The liver is the main detoxifying organ where drug metabolism takes place. However, the enzymes that act on them to deactivate them once they enter the human body do not always give rise to inactive products (metabolites), but sometimes generate other compounds that still have pharmacological activity and require successive metabolism steps for their complete deactivation. This is the case for codeine and morphine, which are present in laudanum extracted naturally from opium poppy

Interestingly, despite living in a globalized world, the availability of different opioid formulations varies from country to country. In addition, drugs are still prescribed according to the intensity or duration of pain rather than according to the pathophysiological mechanism that causes it. Perhaps because it is difficult to understand pain as a disease with its own pathophysiology, it is more often seen solely as a *symptom*. Resistance to pharmacological treatment of pain may explain why it is still undertreated and without good relief.

## Opioid Crisis in America: A Story of Epidemic and Shortage

In 1986, the World Health Organization (WHO) presented a three-step analgesic ladder to serve as a guide in the management of cancer pain. This ladder established that the treatment of different intensities of pain should be in accordance with the efficacy of the pain relief drug. For mild pain, the use of nonopioid analgesics was recommended—mostly nonsteroidal anti-inflammatory drugs such as aspirin, ibuprofen or, alternatively, paracetamol. In their absence (or in combination with them), the use of adjuvant drugs was advised, that is, drugs that are not themselves analgesics but that improve or facilitate pain relief, such as corticos-teroids, muscle relaxants, antiepileptics or antidepressants. Nonopioid analgesics have an analgesic ceiling[5] and can be toxic at high doses or after continuous intake. Therefore, if the pain intensifies moderately, weak opioids could be used instead of these drugs or in combination with them in a second step. The objective would be twofold: on the one hand, it would reduce the intake (dose size or number of ingestions) of the aforementioned drugs; on the other hand, it would address the pain by means of a different, additional mechanism of action. These weak opioid drugs would still have an analgesic ceiling, meaning that at a maximum dose there would be pain that they would fail to relieve. However, the typical side effects of these narcotics (drowsiness, mental fog, nausea, constipation, slowed breathing, dependence, addiction, or tolerance) would be less than those of other types of opioids. Examples of these would be codeine and tramadol. A third and final step would be treatment for severe pain, where strong opioids such as morphine, oxy-codone and fentanyl could be used alone or in combination with nonopioid drugs. These drugs would have virtually no analgesic ceiling, but would have serious side effects if administered continuously or at high doses. The idea is that if the drug circulates throughout the body once ingested, and is so effective in eliminating pain, it may also be very effective in producing other types of effects by binding to various targets and locations.

Although originally created to improve the therapeutic approach in the treat-ment of pain induced exclusively by cancer, its use—or alternatively modifications of it[6]—soon spread to the management of any type of pain (Fig. 3). In fact, cur-rently, the use of the main analgesics (nonsteroidal anti-inflammatory drugs and opioids) is generally conditioned by the duration and intensity of pain, when it should truly depend on the mechanism that generates that type of pain.[7] Therefore, although opioid analgesics have shown high efficacy in the treatment of chronic

---

[5] In pharmacology, the *analgesic ceiling* is the maximum dose above which it is not possible to increase the analgesic efficacy of a drug. Thus, a drug with a low analgesic ceiling will be able to treat only mild pain. If the pain becomes more intense, the drug will not be able to alleviate the pain no matter how much its intake is increased. In other words, the mechanism by which the drug acts on pain transmission is not sufficient to silence neuronal activity when the incoming infor-mation is too great. Increasing its intake, however, can increase the number or intensity of adverse effects, since, with more drug in the body, it can continue to exert its action in other places - causing gastrointestinal, hepatic, or renal damage, for example.

[6] Currently there is a preference for an *analgesic elevator*, which would allow greater flexibility as it would not be an essential requirement to go through each of the steps of the *ladder* in an orderly manner, but could be accessed immediately to any of the levels and even go back and forth from one floor to another, depending on the level of pain experienced by the patient. This elevator would basically consist of three or four floors depending on the degree of pain: mild, moderate, severe or unbearable.

[7] The symptoms and mechanisms of the pathophysiology of pain differ widely due to the often multifactorial nature of pain; no two pains are the same. Just as not all venereal diseases (syphilis, gonorrhea, chlamydia, or AIDS among others) are treated in the same way, even if they present the same virulence or intensity. Each type of pain may involve different tissues, receptors, mechanisms, pathways and signaling fibers. There are different types of pain with different characteristics:

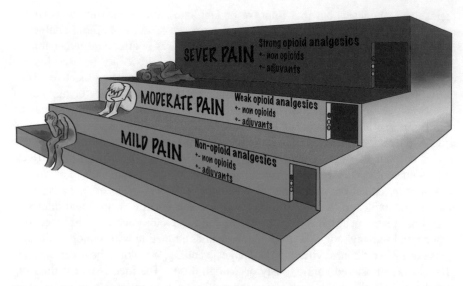

**Fig. 3** Representation of the *analgesic ladder* and the *analgesic elevator* against cancer in **adults**. From the base of the ladder, each step represents a degree or intensity of pain in ascending order, together with the different types or families of drugs recommended for correct pain relief

cancer pain, in many cases their role in the treatment of chronic noncancer pain still depends on a methodology of *trial and error* that is sometimes unsatisfactory. Precisely because it is chronic pain, there is a serious risk of arousing dependence, addiction, and analgesic tolerance with continued consumption, coupled with constipation.[8] Although the endogenous opioid system is currently considered the most effective pharmacological target to produce analgesia, not all opioids are suitable for relieving any type of pain. This is due to the complex pain transmission network but also to the existence of an opioid analgesic ceiling for certain types of pain.

Precisely coinciding with this extension of the use of the WHO analgesic ladder for the treatment of noncancer pain, in recent years, the consumption of opioids has increased significantly, particularly in countries with higher incomes. This opioid overuse, coupled with a high addictiveness, has highlighted a major social issue in healthcare. It has also generated an inordinate caution (which could mean a deficit in the treatment of pain relief for many patients), since its consumption in the

---

dull, burning, oppressive, throbbing, stabbing, sharp, piercing, piercing, tearing, crushing, annoying, intense, strong, intermittent, constant, with a feeling of cramping, localized, diffuse, sweet or debilitating, annoying, to pulses, and spontaneous to movement, touch, cold, heat, air currents. Sometimes, and more so in the case of analgesics, it is forgotten that the drugs were designed to act on specific targets and signaling pathways.

[8] Between 40 and 90% of patients on opioids suffer from constipation, according to the *Handbook of Constipation in Pain Patients* by a panel of European experts (United European Gastroenterol J. 2019;7:7–20).

world varies enormously according to region: 80% takes place in the USA, just 5% of the world's population—if additionally considering Canada, Western Europe and Oceania, these numbers rise to 92% of global opioid consumption—17% of humanity. Surveys by the US National Safety Council reveal that the majority of US doctors exceed recommended doses. In Europe, however, stricter national pharmaceutical regulatory, health care systems, and cultural differences, make abuse there, and other parts of the world less of a problem; these diverse factors play a major role in how opioids are advertised, sold and prescribed in different countries. In North America, the health care system is much more industry-like than in Europe, where pharmaceutical regulation is centralized, and limits how much doctors can prescribe and spend. In Europe, unlike the US, opioids are generally prescribed by specialists, rather than by primary care physicians, and it is forbidden to advertise opioids directly to patients. The motto in the US could be: *Got money, no pain.* If opioids are the most generally effective painkillers for the relief of any intensity of pain, and you are paying a significant amount to maintain private health insurance, it is normal that you do not want to suffer any pain at all. This makes the US the country with the most opioid overdose deaths in the world, between a quarter and a third of the total (including recreational abuse deaths from heroin). Moreover, approximately 3 percent of the US population abuses opioids, compared to 0.4 percent of the European population. Even Germany, the largest opioid-using country in Europe, prescribes half as much opioids as the US. In the US and Canada, there is clear overuse, and it is common to use opioids for an average of 90 days after surgery, regardless of whether it is minor (ingrown toenail, epidermal cyst, papilloma and wart resection) or major surgery. Therefore, drawing conclusions or conducting standardized approaches on a global level entails certain biases.[9]

However, the overprescription of opioid painkillers in the US has been a global trending topic in recent years, leading to an epidemic of addiction and overdose whose spread to other parts of the world has long been viewed with fear and concern, especially in Europe. The overuse of opioid medication in the US has led to an epidemic level of overdose deaths. In fact, US life expectancy dropped for the first time in 2015, in part and as a major contributor, due to the opioid epidemic.[10]

---

[9] As an example, oxycodone, the most prescribed oral opioid in the world—actually in North America—was first synthesized in 1916 in Germany. It reached half the world in less than ten years, and today is the most commonly used major opioid in the US. In the US, oxycodone became fashionable in the 90 s, and was a substance of addiction for the elite because of its high cost—between $20 and $40 per tablet. However, its commercialization in Spain dates only from 2004. This illustrates that different historical, legal, and cultural contexts, as well as political, economic and social contexts can sometimes create a disproportionate global alarm.

[10] Epidemic is a disease, illness, or harm that spreads over a period of time in a place, simultaneously attacking a large number of people. The indiscriminate use of opioids generated many overdose deaths because they produce respiratory depression, hallucinations, suicides, withdrawal syndrome, and increases in the number of domestic and traffic accidents.

Thus, in 2017 and again in 2018, under two different presidents, the opioid epidemic was declared a national public health emergency in the US. During this same epidemic crisis, there has been, paradoxically, another kind of crisis: a shortage of injectable opioids in many US hospitals. The US government tried to solve the epidemic crisis by implementing annual production restrictions on pharmaceuticals; however, almost two-thirds of injectable opioids in the US are produced by a single pharmaceutical company, Pfizer. After receiving a warning in June 2017 by the US Food and Drug Administration (FDA) for noncompliance with established good manufacturing practices, Pfizer began to reduce its production. These measures did not manage to significantly reduce the number of overdose deaths, but they did limit the availability of intravenous opioids for patients who truly needed them, including the three most commonly consumed opioids by the parenteral route: morphine, hydromorphone, and fentanyl. This shortage was exacerbated, in part, by a hurricane in Puerto Rico, a major source of pharmaceutical manufacturing. In the wake of this, many in the US have wondered whether they should expand efforts to source drugs from other countries, rather than have the pharmaceutical market controlled by a small group of companies. Various measures have resulted in the reduction of opioid prescriptions in the US, and there has been a change in the amount and type of opioids used. However, drug shortages, which tend to be relatively long-lasting (see the FDA's drug shortage tracking page: https://www.accessdata.fda.gov/scripts/drugshortages/), are occurring with increased frequency. Given the political, economic, and social contexts, history seems doomed to repeat itself.

There is no more effective drug today than opioids, yet only approximately 17% of the world has access to them. Given that opioid use varies tremendously around the world, the WHO has stated that standard guidelines and regulations are needed to avoid common misconceptions. However, while there is a tendency to globally extrapolate the interactions, adverse effects, and complications that most commonly occur in the US, perhaps these should be viewed more as a *snapshot* of the experiences specific to the US—taking them as a cautionary tale of situations to avoid, instead of guidelines to be followed in other parts of the world.[11]

## Opioid Crisis in Africa: The Tale of the African Peach Tree that Takes Away the Pain

In 1962, a German laboratory specializing in the development of drugs for the treatment of pain managed to combine an opioid agonist, a serotonin reuptake inhibitor, and a norepinephrine reuptake inhibitor for the first time in a single

---

[11] The Spanish Agency of Medicines and Health Products (AEMPS) has recently announced, however, that the consumption of opioids in Spain between 2008 and 2015 increased by 83.6%. Could we be facing a turning point?

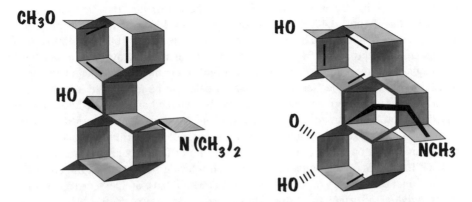

**Fig. 4** **Comparison between the structures of tramadol** (left, bicyclic) **and morphine** (right, tricyclic). The structure of tramadol resembles an unfinished morphine molecule, in which the central component is preserved. The fact that it is "unfinished" allows it to be more flexible and to adapt to other receptors

molecule[12]; each of them is a neurotransmitter involved in the regulation of neuronal activity and, therefore, three of the most efficient mechanisms in pain relief. The drug did not come on the market in Germany until 15 years later, 33 years later in the United States, and 38 years later in Spain. It took 15 years to put the drug on the market! Today, it is one of the most popular drugs in the treatment and relief of pain worldwide, and despite being a narcotic, its risk of drug dependency is somewhat lower than others of the same type, perhaps due to a weaker affinity for the MOR opioid receptor (Fig. 4). The case of *tramadol* is a clear example of the brilliance of the human mind, and the triumph of an idea: circumvent the designs of nature by designing a molecular "Swiss army knife" out of thin air. Currently, its use is restricted to moderate to severe pain, alone or in association with other analgesics with a different mechanism of action, such as paracetamol.

African traditions have generally been transmitted through oral culture in the form of colorful tales and moral proverbs; folk knowledge of plants is no exception. All ethnologists working in Central Africa will have observed that the indigenous people—especially tribal healers—possess a deep knowledge of their natural environment.

In 2013, French researchers discovered high levels of tramadol in the bark and inside the roots of African peach trees in Cameroon. Concentrations ranging from

---

[12] Serotonin is a central neurotransmitter that plays an important role in mood, anxiety, sleep, pain and eating behavior. Serotonin reuptake inhibitors prolong the time serotonin remains in the intersynaptic space, prolonging its effect on mood and chronic pain. Norepinephrine, another major neurotransmitter in the central nervous system, has, among other functions, the ability to regulate alertness and control chronic pain along with serotonin. Norepinephrine reuptake inhibitors prolong the time that norepinephrine remains in the intersynaptic space, exerting a negative effect on the release of more norepinephrine vesicles from the presynaptic neuron.

0.4 to 3.9% of its dry weight were found in bark extracts. If a tree can easily have 100 kg of root, 0.4% would be equivalent to 400 g of tramadol per root. Taking into account that tablets are synthesized for 50 or 100 mg, 4000–8000 tablets could be produced from one tree. In light of this discovery, it is not hard to imagine pharmaceutical competitors rubbing their hands together at the idea that this would cause the German company's patent to be reduced to a dead letter. After 40 years, the German researchers who had created and patented a molecule based on morphine learned that they had invented nothing. Unbeknownst to them, tramadol already existed in nature. It is the same molecule artificially synthesized by humans, with no more or less radicals and no different bonds. It was a major setback for the German company. It was such relevant news that it jumped to the international press. From then on, the Africans would have an inexhaustible source of *designation of origin* tramadol. These peach trees, also known by the name Guinea peach, *Sarcocephalus latifolius*, or *Nauclea latifolia*,[13] have been widely used in traditional medicine in different regions of Africa to treat a variety of diseases, such as malaria, epilepsy, diabetes, various infections, and pain. In fact, in Cameroon itself, their leaves, sap, and roots have often been used for febrile states, abdominal pains, jaundice, malaria crises, or infantile convulsions.

A year later, in 2014, the story took a turn. A German group from the University of Dortmund suggested that the tramadol found in the roots of the peach tree could be explained by a case of human contamination derived from a misuse of veterinary practices. This opinion was initially based on the absence (or traces at most) of the drug in some peach trees, and on the finding of traces of tramadol and three of its metabolites usually generated in mammals (O-desmethyltramadol, N-desmethyltramadol and 4-hydroxycyclohexyltramadol) in roots of five plant species and in soil collected over a wide area. The French group, for their part, argued that it was difficult to explain the high concentration found in samples of root bark and internal plant tissue obtained in a reserve where human activity and livestock grazing are prohibited. In addition, it was difficult to envisage that the plants had the capacity to uptake tramadol from the soil, where it was only present at levels of $10^{-8}$ to $10^{-9}$ g/L, and that it could concentrate in the roots at a level as high as 0.4%. The war was on! The problem lies in the fact that a route to explain its biosynthesis (the identification of enzymes and genes by which the new tramadol was accomplished in nature) had not been mapped. Furthermore,

---

[13] All plants that populate the Earth have different concentrations of organic compounds known as phytochemicals, mainly: terpenoids (carotenoids, saponins), alkaloids (such as morphine) and various nitrogenous compounds, organosulfur compounds, and phenolic derivatives (flavonoids, lignans, stilbenes, tannins, coumarins, phenolic acids). These compounds give the characteristic color, aroma, or flavor to the plants and their fruits, but they also have different antioxidant, analgesic, antitumor, or cardioprotective properties, among others. Indoloquinolicidins are a type of alkaloids mostly restricted to West African plants and in this respect, most of the indoloquinolicidins present in the African peach tree are naucleamides, hence the name of the genus of the species: *Nauclea*.

the original 2013 study validated the traditional use of *N. latifolia* root bark decoctions but not other parts of this plant for pain treatment. Only the roots contained it; no tramadol could be detected in any of the aerial parts of the plant (trunk, branches, leaves), and the 0.4 percent finding did not detail how many plants had been examined.

The field remained open until it was discovered in 2016 that farmers in the area were treating their cows with tramadol and, as is often the case, antibiotics and painkillers were given in large quantities. The cows would gather in the shade of trees and urinate. In fact, today, some farmers have been known to give tramadol to cattle to make them able to plough large tracts of land at once, or to make them able to walk long distances in search of grass even in the sweltering heat of such latitudes. And not just cattle: in the last decade, the African continent seems to have seen an increase in the recreational use of drugs such as tramadol for producing euphoric effects similar to heroin, another opioid. This is so common that tramadol users are called *Tramore* in Ghana (the name of the drug there is *Tramal*); the problem also occurs in Gabon, Egypt and Nigeria. This is probably because, unlike other opioids, such as methadone and fentanyl, tramadol is not subject to international regulation; hence, it is cheap and easy to acquire. Statements have been collected from people claiming that tramadol helped them work long hours by alleviating the pain caused by hard physical labor. In fact, as various testimonies from local people interviewed indicate, using tramadol for work purposes has created a problem for them: dependence, and when it is not taken the characteristic withdrawal symptoms such as nausea and weakness soon appear. This is why they are forced to take it also on days when they do not have to work. Today, consuming tramadol with energy drinks, or dissolved in fruit juices (to camouflage its bitter taste) simply for pleasure at social events, is seen by many as a socially accepted practice and a more discreet way to get high, in contrast to the characteristic smell of cannabis smoke.[14]

In parallel to the American opioid crisis, in recent years this has led to a controversial news story that is not as often discussed: almost 90% of all confiscated opioid drugs, mainly tramadol, are seized in African countries, mostly in West, Central, and North Africa. All indications are that international organized crime syndicates smuggle these drugs across African borders due to a lack of international regulation that would facilitate truly cheap tramadol production. In fact, China and India have become leading exporters, behind the German company producing the original drug Grünenthal, which has been losing considerable rights of its patent. Smuggling from these countries was initially done through Egypt, but when regulations at the borders were tightened, tramadol was smuggled and distributed on the continent via Libya. Companies producing generic tramadol in India and China have in recent years increased the dosage of the tablet to 250 mg,

---

[14] Another long-established use of tramadol on the mainland would not be related to pain, but to sexual intercourse. It is thought that the serotonergic antagonism of tramadol may delay ejaculation.

instead of the recommended dosage of 50 or 100 mg. Why is there not more international scrutiny from drug regulatory agencies? The main concern is that if international control is exercised over the medical availability of tramadol, Europe could suffer from a shortage, with developing countries being especially hard hit. Access to strong opioids is greatest in high-income countries; virtually no strong opioids are available in developing countries. This could result in patients not being adequately treated for pain.

Currently, available analgesics fail to fully treat all types of pain; more effective therapies are needed. While new drugs are being developed, combination therapy using lower doses of drugs with different mechanisms of action seems to be a good short to medium term solution. Perhaps the best medicine is preventive medicine imparted not to the specialist or the primary care physician but to the general population—to inform and educate. If patients have these resources, they will have more resources to deal with their illness, and will be able to interpret their experience not from fear, religion, or magic ("how does the pill know which tooth hurts me?"), but from information and understanding. This will result in better management of pain and of their life. We are witnessing a great technological advance that has generated a culture of the "easy and fast"; however, in a slow health system where a consultation with a specialist can take months, coupled with pain sometimes itself complex, we see that pain management is not so easy or so fast. The situation is further complicated when patients, due to lack of knowledge, experiences their disease from fear.

It would be worthwhile to reflect on how drugs are studied and used in Europe. It would be beneficial to take a more holistic approach—one that integrates the different disciplines, is not burdened by a strict comparison to other countries with very different political, economic and social contexts, and an even more restricted system of pharmaceutical regulation and access to social security. Maybe then we will see that sometimes, the grass is not always greener on the other side of the fence.

## Bibliography

Alfilalo M, Stegmann J-U, Upmalis D (2010) Tapentadol immediate release: a new treatment option for acute pain management. J Pain Res 3:1–9

Avouac J, Gossec L, Dougados M (2007) Efficacy and safety of opioids for osteoarthritis: a meta-analysis of randomized controlled trials. Osteoarthritis Cartilage 15(8):957–965

Ball K, Bouzom F, Scherrmann JM, Walther B, Declèves X (2012) Development of a physiologically based pharmacokinetic model for the rat central nervous system and determination of an in vitro-in vivo scaling methodology for the blood-brain barrier permeability of two transporter substrates, morphine and oxycodone. J Pharm Sci 101(11):4277–4292

Bee LA, Bannister K, Rahman W, Dickenson AH (2010) M-opioid and noradrenergic $\alpha$2-adrenoceptor contributions to the effects of tapentadol on spinal electrophysiological measures of nociception in nerve-injured rats. Pain 152(1):131–139

Boström E, Simonsson USH, Hammarlund-Udenaes M (2006) In vivo blood-brain barrier transport of oxycodone in the rat: indications for active influx and implications for pharmacokinetics/pharmacodynamics. Drug Metab Dispos 34(9):1624–1631

Boucherle B, Haudecoeur R, Queiroz EF, De Waard M, Wolfender JL, Robins RJ, Boumendjel A (2016) Nauclea latifolia: biological activity and alkaloid phytochemistry of a West African tree. Nat Prod Rep 33(9):1034–1043

Boumendjel A, Sotoing Taïwe G, Ngo Bum E, Chabrol T, Beney C, Sinniger V, Haudecoeur R, Marcourt L, Challal S, Ferreira Queiroz E, Souard F, Le Borgne M, Lomberget T, Depaulis A, Lavaud C, Robins R, Wolfender JL, Bonaz B, De Waard M (2013) Occurrence of the synthetic analgesic tramadol in an African medicinal plant. Angew Chem Int Ed Engl 52(45):11780–11784

Bruera E (2018) Parenteral opioid shortage - Treating pain during the opioid-overdose epidemic. N Engl J Med 379(7):601–603

Brummett CM, Waljee JF, Goesling J, Moser S, Lin P, Englesbe MJ, Bohnert ASB, Kheterpal S, Nallamonthu BK (2017) New persistent opioid use after minor and major surgical procedures in US adults. JAMA Surg 152(6):e170504

Cann C, Curran J, Milner T, Ho B (2002) Unwanted effects of morphine-6-glucoronide and morphine. Anaesthesia 57:1200–1203

Drahl C (2014) Tramadol's newfound natural product status in doubt. Chem Eng News 92(39):34–5. Available at: https://cen.acs.org/articles/92/i39/Tramadols-Newfound-Natural-Product-Status.html

Escobar Izquierdo A, Gómez González B (2008) Blood–brain barrier. Neurobiology, clinical implications and effects of stress on its development. Rev Mex Neurosci 9(5):395–405

Finnerup NB, Attal N, Haroutounian S, McNicol E, Baron R, Dworkin RH, Gilron I, Haanpää M, Hansson P, Jensen TS, Kamerman PR, Lund K, Moore A, Raja SN, Rice AS, Rowbotham M, Sena E, Siddall P, Smith BH, Wallace M (2015) Pharmacotherapy for neuropathic pain in adults: a systematic review and meta-analysis. Lancet Neurol 14(2):162–173

Freund M, Speyer E (1917) Über die umwandlung von thebain in oxycodeinon und dessen derivate. J prakt Chem 94(1):135–178. https://doi.org/10.1002/prac.19160940112

Freund W, Speyer E (1924) Product of reduction of oxycodeinon and process of preparing the same. United States Patent Office US1479293A

Furlan AD, Sandoval JA, Mailis-Gagnon A, Tunks E (2006) Opioids for chronic noncancer pain: a meta-analysis of effectiveness and side effects. CMAJ 174(11):1589–1594

Gálvez R, Ruiz S, Romero J (2006) Proposal of a new analgesic ladder for neuropathic pain. Rev Soc Esp Dolor 13(6):377–80. Available at: http://scielo.isciii.es/scielo.php?script=sci_arttext&pid=S1134-80462006000600001&lng=es&nrm=iso

Garcia MM, Goicoechea C, Avellanal M, Traseira S, Martín MI, Sánchez-Robles EM (2019) Comparison of the antinociceptive profiles of morphine and oxycodone in two models of inflammatory and osteoarthritic pain in rat. Eur J Pharmacol 854:109–118

Gaskell H, Derry S, Stannard C, Moore RA (2016) Oxycodone for neuropathic pain in adults. Cochrane Database Syst Rev. 2016;7:CD010692

Godfraind T (2010) About traditional medicine in Central Africa. Rev Quest Sci 181(3):341–371

Gretton SK, Droney J (2014) Splice variation of the mu-opioid receptor and its effect on the action of opioids. Br J Pain 8(4):133–138

Gyamfi Asiedu K (2018) There is an opioid abuse problem unfolding in African cities and it is not getting the attention it needs. Science students. Quartz Africa. Available at: https://qz.com/africa/1223167/opioid-crisis-China-india-tramadol-flood-african-cities-in-ghana-nigeria-egypt-gabon/

Jeffrey P, Summerfield S (2010) Assessment of the blood-brain barrier in CNS drug discovery. Neurobiol Dis 37(1):33–37

Kneip C, Terlinden R, Beier H, Chen G (2008) Investigations into the drug-drug interaction potential of tapentadol in human liver microsomes and fresh human hepatocytes. Drug Metab Lett 2(1):67–75

Koenig KL (2018) The opioid crisis in America: too much, too little, too late. West J Emerg Med 19(3):557–558

Kotecha MK, Sites BD (2013) Pain policy and abuse of prescription opioids in the USA: a cautionary tale for Europe. Anaesthesia 68(12):1210–1215

Kusari S, Tatsimo SJ, Zühlke S, Spiteller M (2016) Synthetic origin of tramadol in the environment. Angew Chem Int Ed Engl 55(1):240–243

Lorenzo P, Moreno A, Lisazoain I, Leza JC, Velázquez (2008) Basic and clinical pharmacology, 18thEd. Médica Panamericana. Madrid. Martín MI, Goicoechea C. pp 213–27

Lauretti GR, Oliveira GM, Pereira NL (2003) Comparison of sustained-release morphine with sustained-release oxycodone in advanced cancer patients. Br J Cancer 89(11):2027–2030

Lemberg K, Kontinen VK, Viljakka K, Kylänlahti I, Yli-Kauhaluoma J, Kalso E (2006) Morphine, oxycodone, methadone and its enantiomers in different models of nociception in the rat. Anesth Analg 102(6):1768–1774

Lemonde.fr [Internet]. Un arbre africain sécrète un antidouleur; c2016. Accessed 14 Jan 2019. Available from: https://www.lemonde.fr/medecine/video/2016/01/27/unarbre-africain-secrete-unantidouleur_4854385_1650718.html

Lenz H, Sandvik L, Qvigstad E, Bjerkelund CE, Raeder J (2009) A comparison of intravenous oxycodone and intravenous morphine in patient-controlled postoperative analgesia after laparoscopic hysterectomy. Anesth Analg 109(4):1279–1283

Löscher W, Potschka H (2005) Blood-brain barrier active efflux transporters: ATP-binding cassette gene family. NeuroRx 2(1):86–98

McCabe SE, West BT, Boyd CJ (2013) Medical use, medical misuse, and nonmedical use of prescription opioids: results from a longitudinal study. Pain 154(5):708–713

Merab E (2019) Alarm raised as more Kenyans get hooked on prescription medicines. Nation; c 2018. Accessed 14 Jan 2019. Available from: https://www.nation.co.ke/newsplex/opioids/271 8262-4700848-tm1dqt/index.html

Moore KA, Ramcharitar V, Levine B, Fowler D (2003) Tentative identification of novel oxycodone metabolites in human urine. J Anal Toxicol 27(6):346–352

Nakamura A, Hasegawa M, Minami K, Kanbara T, Tomii T, Nishiyori A, Narita M, Suzuki T, Kato A (2013) Differential activation of the $\mu$-opioid receptor by oxycodone and morphine in pain-related brain regions in a bone cancer pain model. Br J Pharmacol 168(2):375–388

Narita M, Khotib J, Suzuki M, Ozaki S, Yajima Y, Suzuki T (2003) Heterologous mu-opioid receptor adaptation by repeated stimulation of kappa-opioid receptor: upregulation of G-protein activation and antinociception. J Neurochem 85(5):1171–1179

Narita M, Nakamura A, Ozaki M, Imai S, Miyoshi K, Suzuki M, Suzuki T (2008) Comparative pharmacological profiles of morphine and oxycodone under a neuropathic pain-like state in mice: evidence for less sensitivity to morphine. Neuropsycopharmacology. 33(5):1097–1112

Nau JY (2013) L'arbre à tramadol existe: il prend racine en Afrique. Rev Med Suisse 9(401):1862–1863

Neogi T (2013) The epidemiology and impact of pain in osteoarthritis. Osteoarthritis Cartilage 21(9):1145–1153

Nilsen E (2019) Why it is so much easier to get an opioid prescription in the US than in Europe or Japan. (Internet). Vox; c2017. Accessed 14 Jan 2019. Available from: https://www.vox.com/policy-and-politics/2017/8/8/16049952/opioid-prescription-us-europe-japan

O'Brien T, Christup LL, Drewes AM, Fallon MT, Kress HG, McQuay HJ, Mikus G, Morlion BJ, Perez-Cajaraville J, Pogatzki-Zahn E, Varrassi G, Wells JCD (2017) European Pain Federation position paper on appropriate opioid use in chronic pain management. Eur J Pain 21(1):3–19

Okura T, Hattori A, Takano Y, Sato T (2008) Involvement of the pyrilamine transporter, a putative organic cation transporter, in blood-brain barrier transport of oxycodone. Drug Metab Dispos 36(10):2005–2013

Okura T, Higuchi K, Deguchi Y (2015) The blood-brain barrier transport mechanism controlling analgesic effects of opioid drugs in CNS. Yakugaku Zasshi 135(5):697–702

Ossipov MH, Porreca F (2005) Challenges in the development of novel treatment strategies for neuropathic pain. NeuroRx 2(4):650–661

Oxycodone KE (2005) J Pain Symptom Manage 29(5 Suppl):S47-56

Park JH, Lee C, Shin Y, An JH, Ban JS, Lee JH (2015) Comparison of oxycodone and fentanyl for postoperative patient-controlled analgesia after laparoscopic gynecological surgery. Korean J Anesthesiol 68(2):153–158

Pergolizzi J, Böger RH, Budd K, Dahan A, Erdine S, Hans G, Kress H-G, Langford R, Likar R, Raffa RB, Sacerdote P (2008) Opioids and the management of chronic severe pain in elderly individuals: consensus statement of an international expert panel with focus on the six clinically most often used world health organization step III opioids (buprenorphine, fentanyl, hydromorphone, methadone, morphine, oxycodone). Pain Pract 8(4):287–313

Plazier M, Ost J, Stassijns G, De Ridder D, Vanneste S (2015) Pain characteristics in fibromyalgia: understanding the multiple dimensions of pain. Clin Rheumatol 34(4):775–783

Ross FB, Smith MT (1997) The intrinsic antinociceptive effects of oxycodone appear to be kappa-opioid receptor mediated. Pain 73(2):151–157

Roth SH, Fleischmann RM, Burch FX, Dietz F, Bockow B, Rapoport RJ, Rutstein J, Lacouture PG (2000) Around-the-clock, controlled-release oxycodone therapy for osteoarthritis-related pain. Arch Intern Med 160(6):853–860

Salm-Reifferscheidt L (2018) Tramadol: Africa's opioid crisis. Lancet 391(10134):1982–1983

Sánchez Bayle M (2017) Public health in Madrid five years after the sustainability plan. 3 Nov 2017. Sinpermiso. Available at: http://www.sinpermiso.info/textos/la-sanidad-publica-en-madrid-cinco-anos-despues-del-plan-desostenibilidad

Sanz OJ (2005) Oxycodone. Rev Soc Esp Dolor 12(8):525–531

Scheck J (2021) Tramadol: the opioid crisis for the rest of the world. (Internet). The Wall Street Journal; c2016. Accessed 22 Jan 2021. Available from: https://www.wsj.com/articles/tramadol-the-opioid-crisis-for-the-rest-of-the-world-1476887401

Schmidt-Hansen M, Bennett MI, Arnold S, Bromham N, Hilgart JS (2015) Oxycodone for cancer-related pain. Cochrane Database Syst Rev (2):CD003870

Schröder W, De Vry J, Tzschentke TM, Jahnel U, Christoph T (2010) Differential contribution of opioid and noradrenergic mechanisms of tapentadol in rat models of nociceptive and neuropathic pain. Eur J Pain 14:814–821

Schröder W, Tzschentke TM, Terlinden R, De Vry J, Jahnel U, Christoph T, Tallarida RJ (2011) Synergistic interaction between the two mechanisms of action of tapentadol in analgesia. J Pharmacol and Experim Ther

Sheather-Reid RB, Cohen ML (1998) Efficacy of analgesics in chronic pain: a series of N-of-1 studies. J Pain Symptom Manag 15(4):244–252

Silvasti M, Rosenberg P, Seppälä T, Svartling N, Pitkänen M (1998) Comparison of analgesic efficacy of oxycodone and morphine in postoperative intravenous patient-controlled analgesia. Acta Anaesthesiol Scand 42(5):576–580

Soumerai TE, Kamdar MM, Chabner BA (2019) The other opioid crisis: just another drug shortage? Oncologist 24(5):574–575

Thibault K, Calvin B, Rivals I, Marchand F, Dubacq S, McMahon SB, Pezet S (2014) Molecular mechanisms underlying the enhanced analgesic effect of oxycodone compared to morphine in chemotherapy-induced neuropathic pain. PlosONE. 9(3):e91297

Thorn DA, Siemian JN, Zhang Y, Li J-X (2015) Anti-hyperalgesic effects of imidazoline $I_2$ receptor ligands in a rat model of inflammatory pain: interactions with oxycodone. Psychopharmacology 232(18):3309–3318

Torres LM, Calderón E, Pernia A, Martínez-Vázquez J, Micó JA (2002) From the stairs to the elevator. Rev Soc Esp Pain 9:289–290

Tuchmal M, Barrett JA, Donevan S, Hedberg TG, Taylor CP (2010) Central sensitization and $Cav\alpha2\delta$ ligands in chronic pain syndromes: pathologic processes and pharmacologic effect. J Pain 11(12):1241–1249

Tzschentke TM, Christoph T, Kögel B, Schiene K, Hennies HH, Englberger W, Haurand M, Jahnel U, Cremers T, Friderichs E, De Vry J (2007) (-)-(1R,2R)-3-(3-Dimethylamino-1-ethyl-2-methyl-propyl)-phenol hydrochloride (Tapentadol HCl): a novel mu-opioid receptor agonist/norepinephrine reuptake inhibitor with broad-spectrum analgesic properties. J Pharmacol Experim Ther 323(1):265–276

Tzschentke TM, Christoph T, Schröder W, Englberger W, De Vry J, Jahnel U, Kögel BY (2011) Tapentadol: with two mechanisms in a single molecule effective against nausea and neuropathic pain. Präklinischer Überblick

Uchitel OD, Di Guilmi MN, Urbano FJ, Gonzalez-Inchauspe C (2011) Acute modulation of calcium currents and synaptic transmission by gabapentinoids. Channels 4(6):490–496

U.S. Food and Drug Administration [Internet]. FDA Drug Shortages: Current and resolved drug shortages and discontinuations reported to FDA; c2021 [accessed January 22, 2021]. Available from: https://www.accessdata.fda.gov/scripts/drugshortages/

Vander Weele CM, Porter-Stransky KA, Mabrouk OS, Lovic V, Singer BF, Kennedy RT, Aragona BJ (2014) Rapid dopamine transmission within the nucleus accumbens: dramatic difference between morphine and oxycodone delivery. Eur J Neurosci 40(7):3041–3054

Vidal MA, Calderón MA, Torres LM (2008) Clinical Effectiveness of oxycodone. The 5 mg dose in the analgesic elevator therapeutic scheme. Rev Soc Esp Dolor 15(3):160–169

Wade WE, Spruill WJ (2010) Tapentadol hydrochloride: a centrally acting oral analgesic. Clin Ther 31(12):2804–2818

Wang B, Downing NL (2019) A crisis within an epidemic: critical opioid shortage in US hospitals. Postgrad Med J 95(1127):515–516

Wogan T (2021) Tramadol found in African soils almost certainly artificial. Chemistry World; c2015. Accessed 22 Jan 2021. Available from: https://www.chemistryworld.com/news/tramadol-found-in-african-soils-almost-certainlyartificial/9102.article

Zacny JP, Lichtor SA (2008) Within-subject comparison of the psychopharmacological profiles of oral oxycodone and oral morphine in nondrug-abusing volunteers. Psychopharmacology 196(1):105–116

**Miguel M. Garcia** holds a degree in Biochemistry from Universidad Autónoma de Madrid (UAM), a master's degree in the Study and Treatment of Pain, and a Ph.D. in Pain Research from Universidad Rey Juan Carlos (URJC). He is currently Assistant Professor and Researcher in the Area of Pharmacology, Nutrition and Bromatology of the Department of Basic Health Sciences at University Rey Juan Carlos. He belongs to the High Performance Research Group in Experimental Pharmacology (PHARMAKOM) and coordinates the Teaching Innovation Group in Diseases and their Treatment (EducaPath) of Universidad Rey Juan Carlos. He is a member of the International Association for the Study of Pain (IASP), the Spanish Pain Society (SED) and the Spanish Society of Pharmacology (SEF). His research work focuses on basic pharmacology in the field of pain, particularly on the role of glial cells and cannabinoid and TLR4 receptors in nociception. As a disseminator, he has experience as a speaker at different science festivals and activities.

**Nancy Antonieta Paniagua Lora** has a degree in medicine (specialty in Surgery) from Universidad Mayor Real y Pontificia de San Francisco Xavier de Chuquisaca (USFX, Bolivia). She completed her medical residency program (MIR) in the specialty of Internal Medicine at the Japanese University Hospital and began her research career studying for a Ph.D. in Pain Research in the Department of Health Sciences at Universidad Rey Juan Carlos (URJC). She belongs to the High Performance Research Group in Experimental Pharmacology (PHARMAKOM) of Universidad Rey Juan Carlos, and is a member of the Illustrious Official College of Physicians of Madrid, the Spanish Pain Society (SED) and the Spanish Society of Pharmacology (SEF). She is currently Assistant Professor and Researcher in the Area of Pharmacology, Nutrition and Bromatology of the Department of Basic Health Sciences of Universidad Rey Juan Carlos, and her research work is focused on the evaluation of sensory alterations (*in vitro* and *in vivo*), particularly those induced by peripheral ischemia, and on the characterization of new compounds with analgesic potential. She has participated in dissemination activities such as *Science Week* and the *European Researchers' Night*.

**Eva Mercado Delgado** has a degree in medicine from Universidad Autónoma de San Luis Potosí (UASLP), a master's degree in the Study and Treatment of Pain and a Ph.D. from the Doctoral

Program in Pain Research at Universidad Rey Juan Carlos (URJC). She currently combines her research work in animal models for the study and treatment of pain in the Area of Pharmacology, Nutrition and Bromatology at Universidad Rey Juan Carlos with her activity as an interventional physician and head of the Pain Management Unit at Hospital Universitario Sanitas Virgen del Mar.

# The Curious Case of the Mutations of Dr. Jekyll and Mr. Hyde

Sergio Muñoz Sánchez©

Telekinesis, super strength, super speed, the ability to fly or control storms or the properties of metals at your whim. Sounds good, does not it? These are just some of the powers that the X-Men, the most famous mutants in the comic book world, have. The origin of the powers displayed by this group of metahumans lies in the presence of the so-called X gene—the mutant gene. Most likely, the case of the X-men is one of the few exceptions in which the words "mutation" and "mutant" have positive connotations in popular culture. The word "mutant" comes from the Latin *mutare*, which simply means to change; however, outside the scientific field, the term mutant is often associated with negative, harmful, or damaging aspects. The mention of the word "mutation" probably brings to mind images of deformed beings, with jarring colors, or extra limbs (type "mutant" in an Internet image search engine and take a look at the first results). If we turn off our imagination, and consider a more realistic situation, we will likely think of a disease. And we would be right. In fact, one or more mutations in our genetic material are behind some of the most serious diseases. However, mutations understood as variations in the sequence of our genes (or more broadly, of any sequence of our DNA, whether or not it contains a gene) are not necessarily negative. While far from being able to confer invisibility or telepathy, mutations are, nevertheless, the most basic substrate

S. Muñoz Sánchez (✉)
DNA Replication Group, Molecular Oncology Programme, Spanish National Cancer Research Centre (CNIO), Madrid, Spain
e-mail: smunoz@cnio.es

© The Author(s), under exclusive license to Springer Nature Switzerland AG 2024
M. M. Garcia (ed.), *Tales of Discovery*, https://doi.org/10.1007/978-3-031-47620-4_9

on which natural selection acts[1] and, therefore, responsible for the fact that each species that populates our planet is perfectly adapted to its environment. Let us see how.

First, let us consider what our genetic material is and how it works. DNA is, so to speak, the instruction book required to build a living being and make it "work" properly. In the sequence of this marvelous molecule lies the precise information that determines how, when, where, and in what quantity the different components of our organism are synthesized. Specifically, DNA or *deoxyribonucleic acid* is made up of two long strands on which two fundamental components called *ribose* and *phosphates* alternate as links in a chain. Each ribose carries, in the inner face of the strand, a compound known as *nitrogenous base*. The nitrogenous bases that make up DNA are *adenine* (A), *thymine* (T), *cytosine* (C), and *guanine* (G), and they are complementary in pairs. This means that the A-T and C-G pairs are capable of chemical bonding. In this way, each ribose-phosphate chain faces its nitrogenous bases inwards, and the bond between them holds the two strands together (Fig. 1). Visualize it as a zipper: the ribose-phosphate chains would be the fabric parts and the nitrogenous bases would be the teeth of the zipper that join them together. This zipper, however, turns on itself, generating the very famous double helix of DNA. This apparently simple molecule encloses a great complexity, since the language of life is precisely articulated around the sequence in which the nitrogenous bases are arranged along the strands of DNA. This sequence is used to build proteins and enzymes that we could identify as molecular "machines", which ultimately carry out works as diverse as digesting food, carrying oxygen to our organs, facilitating the contraction of our muscles, controlling the division of our cells, or transforming some molecules into others. But how are these small molecular devices generated (or *synthesized*, to use scientific jargon)? Simply put, a gene is a particular DNA sequence that encodes the information necessary for the synthesis of a specific protein which, in turn, will carry out one or more cellular functions. In a gene, the nitrogenous bases are "read" in trios called *triplets*, and each triplet codes for an amino acid. These amino acids are the "bricks" that make up proteins, and the order in which they are assembled establishes the form and function of those proteins. Thus, the sequence of genes is precisely read by multiprotein complexes to form new proteins that will perform their different functions. In fact, the coordinated action of hundreds, even thousands of proteins in our organism is necessary to execute very complex

---

[1] Natural selection is the main mechanism by which biological evolution operates. It is the survival and differential reproduction of those individuals with genetic material better adapted to their environment. Natural selection gives those individuals higher survival and reproduction rates, which ultimately leads to the fixation (or selection) of these characteristics in the population, contributing to its evolution. As will be seen throughout the chapter, for natural selection to operate, these genetic characteristics must be heritable and present different variants in the population so that only those that allow the individual a better survival and reproduction rate are randomly selected.

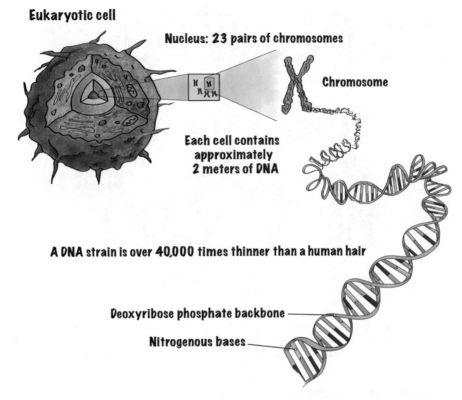

**Eukaryotic cell**

**Nucleus: 23 pairs of chromosomes**

**Chromosome**

**Each cell contains approximately 2 meters of DNA**

**A DNA strain is over 40,000 times thinner than a human hair**

**Deoxyribose phosphate backbone**

**Nitrogenous bases**

**Fig. 1 Schematic representation of the DNA double helix**. In eukaryotic cells (such as human cells), DNA is stored in the nucleus of the cell. Each human cell contains approximately 2 m of DNA organized and condensed into formations called chromosomes. Each DNA strand consists of two opposite strands of ribose-phosphate residues chemically linked by complementary nitrogenous bases (A-T and C-G). The double-stranded thread turns on itself to form a double helix structure

processes (e.g., to control the homeostasis[2] of our different tissues and organs, or to determine physical aspects such as our height, skin, eye, or hair color).

## Mr. Hyde; the Dark Side of Mutations

As stated earlier, strictly speaking, a mutation is nothing more than a change in the sequence of nitrogenous bases of our DNA. Where there was a T, a G appears, for example, and the triplet to which this T belongs changes from TCC to GCC.

---

[2] Homeostasis is known as the set of mechanisms responsible for controlling and maintaining a balance in the different biological systems, buffering the changes produced by external agents, and ensuring their correct functioning. Homeostasis can refer to different systems, from cells to complete organisms.

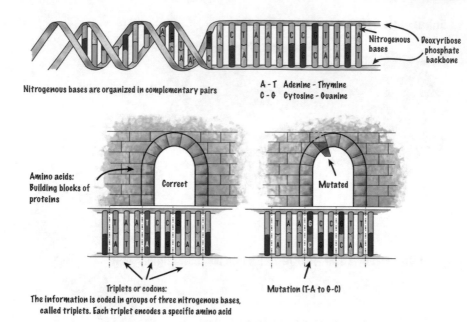

**Fig. 2 Results of a single mutation in the DNA sequence.** The proteins in our body are made up of amino acids linked together in a chain. DNA contains the information needed to build all these proteins. This information is encoded in groups of three nitrogenous bases called triplets or codons. Each triplet encodes the information for the introduction of a specific amino acid into the protein sequence; thus, the sequence of nitrogenous bases arranged in triplets determines the order in which the amino acids combine to form a protein. A change in a base means a change in a triplet and, therefore, the introduction of a wrong amino acid in a specific position of the protein, affecting its structure and function. In the analogy of the wall, the mutation involves the introduction of a faulty brick that affects the structure of the door

The amino acid (the building block of proteins) that it encodes also changes, and with it the sequence of the protein (Fig. 2). Hopefully, the amino acid in question had only an ancillary or secondary role, and the activity of the protein will not be altered. Sometimes, however, the amino acid in question plays an indispensable role in the function of the protein, and its activity is compromised. Here a disease or genetic condition will appear.

Let us continue with the analogy of bricks and masonry to illustrate it better. Imagine a wall. A defective brick placed on one of the battlements at the top will not be a serious problem, but what if the defective brick is part of the arch of the gate? The gate would not close—or worse, the wall could collapse. Now apply this concept to genetics. Phenylketonuria is a *recessive genetic disease* caused when both copies (maternal and paternal)[3] of the gene that codes for the

---

[3] In our cells, the genetic material is duplicated, and we inherit both a maternal copy and a paternal copy; we get two copies of each gene. In cases of *recessive congenital diseases*, one intact copy

enzyme *phenylalanine hydroxylase* are mutated, generating inactive variants of the protein. Although more than 500 mutations have been identified that can cause this disease, the most common is found in the amino acid at position 408, where an *arginine* is replaced by a *tryptophan*, a much larger amino acid. The function of the enzyme phenylalanine hydroxylase is to transform one chemical compound (*phenylalanine*) into another (*tyrosine*). For this, it is necessary that it is associated with groups of 4 units. Position 408 is essential for these associations. When it is mutated, these bonds cannot be established, and the enzyme cannot work. This generates the accumulation of the first compound (phenylalanine), which, in large quantities, is toxic to the central nervous system, causing serious brain damage to newborns carrying the mutant variants of the gene. Fortunately, nowadays, a simple diagnostic test is routinely performed on newborns, which allows early detection of the disease. In addition, the development of the disease in babies carrying the mutations can be prevented by following a specialized diet in which the chemical compound that causes the problem is supplied in very limited quantities. Phenylke-tonuria is a classic example of how a point mutation in a particular gene causes a disease. Such conditions are known as *monogenic diseases*. However, in many cases, it is the combination of multiple mutations in various genes, and/or DNA regulatory sequences that ultimately lead to failure. These diseases are known as *multigenic* due to their complex nature. A clear example of this situation would be cancer.

Cancer is an uncontrolled growth of certain cells in our body; in this case, growth refers to cell division, also called *mitosis*. Cancer cells divide uncontrollably, and different mutations in the DNA allow them to bypass the control mechanisms established by our own genes. These anarchic divisions end up generating a tumor and with it the disease. The term "cancer" was coined in classical Greece, and comes from the word *karkínos*, which means crab. This word, and its Latin derivative *cancer*, were used to describe both the tumors that today are included within the term "cancer", as well as other types of ulcerous pathologies that have nothing to do with this disease. The term was coined because of the external appearance of the tumors, especially those of the breast: a hard, reddish mass from which a small network of blood vessels emerges, resembling, respectively, the body and legs of a crab. This is why the disease has the same name as the constellation Cancer, the mythological crab that Hercules faced during one of his 12 labors. In Spain, breast cancer is also known as "*zaratan*", a word that comes from the Arabic "*saratan*", which also means crab, and which medieval Muslim doctors used to refer to this pathology.

Returning to the subject at hand: what kind of mutations cause cancer? As we mentioned before, our genetic material—our instruction book—imposes multiple

---

of the gene is suficient to produce functional proteins, so the disease would only appear if both copies are mutated (*recessive variants* of the gene). In cases of *dominant congenital diseases*, it is sufficient that only one of the two copies is mutated (*dominant variant* of the gene) for the disease to appear.

control mechanisms on our cells so that mitosis occurs only when it is necessary (e.g., during embryonic development or during the healing of a wound). Put simply, two types of genes are at the helm of these control mechanisms: *proto-oncogenes* and *tumor suppressor genes*. To use an analogy from one of my genetics professors, the proto-oncogenes would be the accelerator of a car, and the suppressor genes the brake. Proto-oncogenes (more specifically, the proteins they encode) are responsible for starting and executing cell division. In contrast, tumor suppressor genes are in charge of stopping it, establishing a balance that results in correct cell homeostasis. The mutations that alter this balance and give rise to cancer are different in each genetic group. In this context, we can classify the types of mutation into two broad groups: loss-of-function mutations (the device is constantly kept off), and gain-of-function mutations (the device is constantly kept on). You may have already anticipated which type of mutation affects which type of gene. In the case of tumor suppressor genes, the mutations responsible for the genesis of the tumor are loss-of-function mutations, which result in functionless protein variants (as was the case with the mutations responsible for phenylketonuria). The brake does not work, and the car accelerates. If these types of mutations occurred in proto-oncogenes, they would never lead to tumor growth. If the accelerator does not work, the car stops. For the car to pick up speed, the accelerator would have to remain activated, regardless of whether it was operated by the driver. This is exactly what happens when a proto-oncogene mutates. During tumor development, one or more proto-oncogenes undergo gain-of-function mutations, and the proteins they encode remain constantly active. The reasons are very varied. The mutation could result in the loss of a segment of the protein necessary for its inhibition (or "switch-off"), or perhaps in the production of more amount of protein than necessary. These are just a few examples. In either case, the result would be that the function of the protein is altered so that the protein behaves like a device without an off switch. The car accelerates.

Fortunately, our cells do not have just one accelerator and one brake for mitosis; they have hundreds of them. Each accelerator and each brake respond to different types of signals, and on many occasions, the same type of signal activates different accelerators or brakes, so that the control mechanisms of mitosis are usually redundant. In this way, the cell ensures that there is a plan B in case of failure, a safeguard, a safety net that prevent it from falling if the trapeze rope breaks. In fact, the most accepted hypothesis is that at least two alterations (two hyperactivated proto-oncogenes, two inactivated tumor suppressors, or one of each) are needed for the development of the vast majority of known tumor types. In addition, for the disease to develop, other processes beyond mitosis must be altered by additional mutations. To continue dividing, tumor cells must "do" other things: alter the metabolism of fatty acids and sugars (their way of feeding); promote angiogenesis (the generation of blood vessels that carry blood to the tumor); modify the capacity for cell adhesion (which would allow them to detach from the tissue and invade other organs causing metastasis); or modify the immune response (the host's defense against the tumor). Each of these processes has its own control mechanisms, so our body is equipped with plenty of safeguards against tumor development. Note that the word *do* above is in quotation marks. The reason is that cells do not purposely *do* any of these things—they do not have mechanisms to specifically mutate any gene. Rather, these genes mutate randomly, and these mutations are subsequently selected by conferring an advantage, in this case the ability to bypass a control mechanism. This selection process will be better understood below, when we discuss how natural selection acts on DNA mutations.

## Dr. Jekyll and Evolution; the Kind Face of Mutations

At this point, the reader may be thinking that the negative view of mutations is more than deserved. What is the positive aspect of my body accumulating chemical compounds that should not be there, or of my cells dividing haphazardly? There is no positive aspect whatsoever. These mutations and their consequences are one side of the coin. They are the dark side. They are Mr. Hyde. And because every coin has a flip side, there must also be a Dr Jekyll. Let us use another automotive analogy to get to know him. Imagine a Formula 1 World Championship. At the Abu Dhabi Grand Prix, the drivers use dry tires, as rain rarely appears at those latitudes. Suppose the rules allowed only one type of tire to be used for the entire race, the teams would confidently go with dry tires. However, under such a hypothetical rule, which tire would they use in a place with more variable weather? What would they use on a cloudy day at the Spanish Grand Prix at the Circuit de Catalunya? Some would choose dry tires, guessing that it will not rain; others would use wet tires, betting on the opposite case. Those who get the weather right will have a clear advantage, and probably take first places, while the rest may not even finish the race. However, that is just a hypothetical, and in the World Championships the rules allow pit stops where you can change tires to adapt to a changing environment. Mutations offer this possibility to living beings.

Natural selection is the main engine of evolution, choosing the fittest to the detriment of the rest. However, the very concept of "choice" necessarily implies the existence of several options from which to choose. Mutations introduce variation; they occur randomly, and are not directed. Ultraviolet light, oxygen free radicals, various environmental components, or DNA metabolism itself (the expression and maintenance of genes) introduces changes in our genetic material. Most of these changes are corrected by DNA repair mechanisms. This cellular machinery recognizes the damage and changes in the code, and reverses them. To a large extent, DNA repair pathways are partially responsible for diseases such as cancer being much less abundant than they could be. Consider that our DNA, made up of approximately 3.3 billion base pairs, can undergo up to a million lesions in a single day. Although it might seem a small percentage of the total (approximately 0.03%), just one or two of these mutations affecting a proto-oncogene or tumor suppressor gene would be enough to have catastrophic consequences. Hence, the need for these repair mechanisms, the discovery of which deserved the 2015 Nobel Prize in Chemistry for scientists Tomas Lindahl, Paul Modrich, and Aziz Sancar.

As necessary as these mechanisms are they are not infallible, and that is where natural selection come in. Imagine a hypothetical and perfectly idealized case of how natural selection would act on these random mutations. One of these random mutations could affect a gene in such a way that, far from being inactivated, the protein could develop greater efficiency or even gain a new function. That new function could provide some beneficial advantage for the organism, and it would be "chosen" by natural selection. This advantaged individual would probably reproduce more than those less fit, so that the beneficial mutation would be passed on to the next generation and, more importantly, would be passed on in greater proportion than the less favorable variants. The repetition of this process over generations would contribute to selecting for and perpetuating this variant in the population and, therefore, would contribute to the evolution of the species.

Natural selection and evolution run through much more complex channels, but this simplified case gives us an overview of the process. To better illustrate the role that mutations have played (and still play) in the natural history of our planet, we will discuss some concrete examples. We will start with a simple and well-known example that appears in most zoology or ecology textbooks: the case of the *Biston betularia*, also known as the peppered moth. This moth usually has a light gray color with some darker spots that allows it to camouflage itself perfectly in the bark of birch trees covered with lichen, and avoid being seen by its natural predators, the birds. However, there is also a dark gray variety that appeared in the UK in the mid-nineteenth century, coinciding with the Industrial Revolution. The soot produced by the increased and extensive use of coal at this time prevented the growth of lichens, and stained black the tree trunks on which our moth rested. In these conditions, their light grey coloring was clearly ineffective in hiding them from predators, who were quick to spot hem. However, the dark moths that were previously easily spotted when they landed on birch trees were now virtually undetectable. This new characteristic—an advantage—was quickly selected so that the proportion of the dark variant (suitably called *carbonaria*) increased rapidly in the total moth population. Studies in the second half of the nineteenth century in industrial areas of England documented a 98% increase in the carbonaria variety over the common variety. This phenomenon, known as *industrial melanism,* is one of the first recorded cases of natural selection, and has been widely studied since it was first observed. It was not until 2016 that the gene and mutation responsible for the appearance of the dark variant was identified. It is the *cortex* gene. Scientists at the University of Liverpool discovered that the insertion of a large, tandemly repeated, moving DNA sequence (a *transposon*) into the *cortex* gene, resulted in an increase in its expression, and leading to the black coloration typical of this variety.

For our second example, we are going to focus at some of the oldest and simplest organisms that populate our planet: bacteria. Bacteria are *prokaryotic* organisms (meaning that they have no nucleus in which to house their genetic material), with a simple metabolism, and asexual reproduction. This last characteristic is very important for the example at hand. Organisms that reproduce asexually do so by division, after duplication of their DNA, so that offspring are identical copies of their parent (in fact, after division, parent and offspring are virtually indistinguishable). In the face of selective pressure, i.e., an environmental change that harms the species (in our racing car metaphor, a sudden rain when a sunny day was expected), such organisms are at a great disadvantage because they are all the same—true clones. Without different possibilities to choose from, natural selection cannot operate. However, bacteria also have DNA repair mechanisms that are not perfect and leave faults behind: mutations that in a constant environment might seem undesirable to them, but now result in the salvation of the species. The specific example I want to discuss is antibiotic resistance. For a bacterial population, the presence of an antibiotic is a strong selective pressure. An antibiotic is a chemical compound that blocks one or more *enzymes*. Enzymes are a specific group of proteins, the molecular machines we talked about earlier. Without these machines, bacteria die. Quinolones are antibiotics that act by binding to and inhibiting the enzyme DNA gyrase, which is necessary for the DNA duplication that precedes cell division. Sublethal concentrations of quinolones, which would damage the bacterial population but not completely eliminate it, give bacteria enough time to accumulate random mutations in their DNA. If one of these mutations modifies the DNA gyrase gene in such a way that the quinolone binding site is altered, that individual acquires resistance to the antibiotic, and is selected for its evolutionary advantage. While the rest of the bacterial population slowly dies off, the descendants of this mutant individual spread, producing a resistant population in a few generations. Although this is not the only way by which bacteria acquire resistance,[4] this ability underscores the vital importance of avoiding self-medication with antibiotics, and to strictly follow the medication regimen without ending treatment early. Improper or insufficient administration means exposure of bacterial populations to sublethal doses of antibiotics; they are therefore being presented with a strong selective pressure together with the possibility of a chance gene mutation that could confer a resistance that would quickly spread in the population.

This principle was used by the recent Nobel Prize winner in chemistry, Frances H. Arnold, to obtain more active enzyme variants by means of directed evolution. In one of her most important studies, this researcher set out to develop variants of the bacterial enzyme *subtilisin* that would have much higher activity than conventional variants. Furthermore, it was intended that these new variants would be able to work in organic solvents instead of the usual aqueous solvents. These variants

---

[4] The mechanism we are discussing is just a good example to show the importance of point mutations in evolution. The mechanism itself is accessory. Bacteria's most common and efficient way of acquiring resistance is different. Antibiotic resistance is usually determined by the ability of bacteria to share extrachromosomal genetic material called *plasmids*. These plasmids are circular DNA molecules that usually contain resistance genes that encode for enzymes capable of breaking down antibiotics. The doses of antibiotics given during a microbial infection are so high that they often kill the bacteria before they put these genes to work. If these doses are suboptimal or if treatments are stopped early, the bacteria have time to activate and share these genes, contributing to the development of resistance.

would deeply increase the efficiency of several biotechnological processes. The methodology used to obtain them was worthy of such a distinguished prize. Arnold randomly mutated the gene for the enzyme in question, and introduced each variant into a bacterial population. She then put each population under selective pressure to identify the colonies that produced the most active enzymes. Basically, she grew the bacteria on plates combining an organic solvent and milk *casein*, a compound that is hydrolyzed or broken down by subtilisin. Bacteria with more efficient variants of subtilisin grew better and were selected, and then subjected to a new round of mutagenesis and selection. In this elegant way, Arnold obtained enzymes capable of working in organic solvents with an activity up to 256 times higher than that of conventional variants.

For our last example, we will refer to a species that will be much more familiar and recognizable than certain English moths, or bacteria with unpronounceable enzyme names. We will talk about *Homo sapiens sapiens*, and our ability to continue feeding on milk beyond lactation. Regardless of the whether it is good or not to use milk and its derivatives in our usual diet (a subject not discussed here), the ability to continue metabolizing lactose in adulthood is a curious and recent case of mutation followed by natural selection in our own lineage. At the dawn of our species, humans did not differ from other species in our taste for dairy products. Milk, exclusively of maternal origin, was reserved for the young of our species as long as they were breastfed. The main reason is that the gene for *lactase*, the enzyme that digests *lactose*, was inactivated after lactation. With the emergence of livestock and pastoralism, humans began to feed on the meat and milk of their animals. At that time, milk had to be fermented or transformed into other derivatives to be consumed, since the inability to digest lactose generated additional intolerances and health problems. However, approximately 10,000–6000 years ago, a mutation appeared in the people of Northern Europe that prevented the lactase gene from being inactivated in adulthood, allowing its carriers to access a food source that was easy to obtain and very rich in nutrients. This ability to access new resources constituted an adaptive advantage. Additionally, in high latitudes, where daylight hours are scarcer, lactose tolerance was an additional advantage. The lack of sun exposure is a limiting factor in the production of vitamin D, as well as in the fixation of calcium in our bones. At these latitudes, these deficiencies could have been remedied by the direct intake of these compounds from milk. This hypothesis is supported by the fact that, until only 3800 years ago, this mutation was rarely found in the population of southern Europe, as seen from genetic studies based on remains found in prehistoric sites. In any case, it is estimated that among the cultures in which the mutation originated, the advantage it conferred on its carriers could have increased their chances of reproducing by up to 19%.

We could end the story here; it would be attractive enough. However, it has an epilogue that makes it even more interesting. The migrations of the peoples of northern and central Europe spread the mutation throughout the continent, but they did not reach all corners. There is no documented contact between these people and the peoples of other latitudes, such as the African continent, where the ability to digest lactose is also present. So how did this mutation get there? To try to answer this question, a group of scientists from the University of Maryland carried out a genetic study of several current East African populations. These researchers found up to three mutations that were different from the European mutation. The East African mutations reactivated the lactase between 6800 and 2700 years ago in the ancestral inhabitants of this area of the planet. This suggests that the reactivation of the lactase gene was an evolutionary event that appeared several times in the recent history of the evolution of our species. Moreover, each of these events occurred independently in populations that had no contact with each other. In fact, research from fossil remains from the Atapuerca site in Spain shows that the ability to metabolize lactose on the Peninsula was also acquired independently. The selective pressures that facilitated the wide distribution of the aforementioned mutations in Southern Europe and East Africa were different: while in the first case the adaptive advantage would be related to survival in situations of food shortage, in Africa the consumption of milk could have favored the carriers of the mutation not only in situations of famine, but also during periods of long droughts. In the latter case, the "mutant" individuals could see their chances of reproduction multiplied by up to a factor of ten (a superpower even the X-patrol would envy). The lactase phenomenon is a beautiful example of convergent evolution: an event by which two distant, unrelated populations develop the same adaptation to the same or to a different problem, as happened with this particular enzyme. Cases of convergent evolution show how accurate is the strategy developed, as it appears by chance and is selected independently several times during the evolutionary history of the species. It is stimated that up to one-third of the world's population has the ability to digest milk without any problem nowadays. This observation illustrates the success of the reactivation of the lactase gene in our species.

These three stories of evolution show us how natural selection acts at the genetic level, and the importance of variation, change, and ultimately mutations in our DNA for selection to operate. I would like to end this story with a couple of brief ideas so that the concepts presented here do not lead to confusion. The first is that, technically speaking, we are all mutants; we all have some mutation in our organism. There is no standard human DNA sequence, as if it were a Vulgate, so that any variant in a specific position could be considered either standard or mutation. There are just variants that are more or less common than others. In the scientific field, those mutations that do not cause any change in the protein encoded by the gene are considered *genetic polymorphisms* (simply genetic variations), and are called SNPs (single nucleotide polymorphism), while the term *mutation* is reserved for those changes in the DNA sequence that actually affect the function of the encoded protein. So since, strictly speaking we are all mutants, do not hate X-Patrol (at least not because they are mutants).

The other idea I would like to convey is that natural selection and evolution do not operate only at the genetic, cellular or individual level. Several distinguished scientifics, such as Ernst Mayr and Stephen Jay Gould, have shown that evolution is capable of acting on groups or entire species—but that is a story for another occasion.

# Bibliography

Aliu E, Kanungo S, Arnold GL (2018) Amino acid disorders. Ann Transl Med 6(24):471
Blau N (2016) Genetics of phenylketonuria: then and now. Hum Mutat 37(6):508–515
Chen K, Arnold FH (1993) Tuning the activity of an enzyme for unusual environments: sequential random mutagenesis of subtilisin E for catalysis in dimethylformamide. Proc Natl Acad Sci USA 90(12):5618–5622
Erlandsen H, Patch MG, Gamez A, Straub M, Stevens RC (2003) Structural studies on phenylalanine hydroxylase and implications toward understanding and treating phenylketonuria. Pediatrics 112(6 Pt 2):1557–1565
Fàbrega A, Madurga S, Giralt E, Vila J (2009) Mechanism of action of and resistance to quinolones. Microb Biotechnol 2(1):40–61
Hanahan D, Weinberg RA (2011) Hallmarks of cancer: the next generation. Cell 144(5):646–674
Hickman Jr CP (2006) Life: biological principles and zoological science. In: Hickman Jr CP, Roberts LS, Larson A, l'Anson H, Eisenhour DJ (eds) Comprehensive principles of zoology, 13th ed. Madrid: McGraw-Hill/Interamericana de España, S.A.U., 2006. pp 2–23
Mediavilla D (2015) How did we start to drink milk? El País newspaper. (online). https://elpais.com/elpais/2015/09/20/ciencia/1442747482_528167.html
Prescott LM (2004) Antimicrobial chemotherapy. In: Prescott LM, Harley JP, Klein DA (eds) Microbiology, 5th ed. Madrid: McGraw-Hill-Interamericana de España, S.A.U., 2004. pp 849–868
Rivera A (2006) Lactose tolerance in Africa indicates a recent evolution of the human species. El País newspaper. (online). https://elpais.com/diario/2006/12/27/futuro/1167174004_850215.html
The Nobel Prize organization (2019) Press release: the Nobel Prize in chemistry 2015. Nobel Media AB 2019. (online). https://www.nobelprize.org/prizes/chemistry/2015/press-release/
The Nobel Prize organization (2019) Press release: the Nobel Prize in chemistry 2018. Nobel Media AB 2019. (online). https://www.nobelprize.org/prizes/chemistry/2018/press-release/
Tishkoff SA, Reed FA, Ranciaro A, Voight BF, Babbitt CC, Silverman JS, Powell K, Mortensen HM, Hirbo JB, Osman M, Ibrahim M, Omar SA, Lema G, Nyambo TB, Ghori J, Bumpstead S, Pritchard JK, Wray GA, Deloukas P (2007) Convergent adaptation of human lactase persistence in Africa and Europe. Nat Genet 39(1):31–40
Van't Hof AE, Campagne P, Ridgen DJ, Yung CJ, Lingley J, Quail MA, Hall N, Darby AC, Saccheri IJ (2016) The industrial melanism mutation in British peppered moths is a transposable element. Nature 534(7605):102–105

**Sergio Muñoz Sánchez** holds a degree in Biochemistry, a master's degree in Molecular and Cellular Biology and a Ph.D. in Molecular Biosciences from Universidad Autónoma de Madrid-Centro Nacional de Investigaciones Oncológicas (UAM-CNIO). He has conducted several interships at national reference centres (Centro Nacional de Biotecnología, CNB; Centro de Biología Molecular Severo Ochoa, CBMSO; Centro Andaluz de Biología Molecular y Medicina Regenerativa, CABIMER) and currently carries out his research work at the Spanish National Cancer Research Centre (CNIO), where he studies different mechanisms of genomic instability that affect the process of DNA duplication and cell division.

# Evolution, Sex and Pain. Multiple Faces of the Dice

Miguel M. Garcia⊙ and Marta Martín Ruiz⊙

Suppose we are talking about pain. Humans feel pain, but what about a fly or an octopus? How similar are our nociceptive systems? And what about the reproductive system? Humans have two sex chromosomes, but what about other animals? How similar is the number of chromosomes throughout the entire animal kingdom? The two systems are very different in appearance but conserved, in part, throughout evolution. Although different, the essential mechanisms for life overlap and exhibit common patterns between species beyond chance. Adapt or die, in the purest *Darwinian* style.

Let us start with a simple, familiar object: dice. A dice is a polyhedron—usually cubic- with 6 or more faces, which we have all probably used at some time in a board game, or possibly a casino game. Although different in color and shape, all dice have a series of faces, marked with pips, numbers, or symbol, in a way that makes it possible to understand the nature of the dice. In nature, animals all have a series of common "faces" that constitutes the essence of our existence, and without which we could not live. These are, for example, the reproductive, nociceptive or

M. M. Garcia (✉)
Area of Pharmacology, Nutrition and Bromatology, Department of Basic Health Sciences, Unidad Asociada de I+D+i al Instituto de Química Médica (IQM-CSIC), Universidad Rey Juan Carlos, Madrid, Spain
e-mail: miguelangel.garcia@urjc.es

High Performance Experimental Pharmacology Research Group (PHARMAKOM), Universidad Rey Juan Carlos, Alcorcón, Spain

Grupo Multidisciplinar de Investigación y Tratamiento del Dolor (i+DOL), Alcorcón, Spain

M. Martín Ruiz
Laboratory of Molecular Biology, Hospital Universitario General de Villalba, Collado Villalba, Spain

digestive systems, among others. Systems that are perhaps very different at first glance when comparing different species, but essentially conserved over millennia of evolution.

Reproduction is common across species; however, sexual reproduction is unique to multicellular individuals with *eukaryotic-like cells*, and is the main type of reproduction in some groups of organisms, such as animals. Moreover, sex ensures the evolution and diversity of all animal species through two types of specialized cells called *gametes*. Another type of reproduction is asexual (but that is a different story—and much less fun). Similarly, all animals have a nervous system or protosystem capable of making us perceive potentially harmful or dangerous stimuli for our survival or physical integrity and, consequently, we are able to flee from them (whether suffering and pain are universal processes is, to some extent, debatable). This is where the *nociceptive* system comes into play (from the Latin *nocere*, to harm or hurt). In addition, all animals have a digestive system—at a minimum, a mouth and an anus, although sometimes it is the same hole.[1]

Without these systems we could hardly survive, and we would not understand animal life as we presently do.

## dTRP, A Complicated Acronym for Feeling Pain

The fact that we all share the same human condition, means that frequently, the motivations or experiences of others are not too far from our own. Consider, for example, long summer nights when you cannot get to sleep. Mosquitoes, guided by their fine sense of smell, and far from the sun's rays that bother them so much, come out to look for their victims. Surely many of us are familiar with this situation, and more than one of us has swatted in the dark, satisfied when the hand reaches its target. But did the mosquito suffer? Some of us have etched in our brains the image of clams opening when they are steamed, or crabs trying to escape a pot of boiling water. But again, does it hurt? And those who have ever gone fishing likely have pierced a worm with the hook on the end of their line, and watched how it twisted before getting lost under the water. Have you wondered if that hurts?

When we think about the first organisms with which life on Earth presumably began, we are referring to bacteria with motor capacity; that is, organisms capable of moving, turning around, and moving away from a hostile, too hot, or acidic environment by themselves. However, no one will say that this is pain, or that they have a nervous system or a brain, much less that the change of direction is due to a reasoned and voluntary action. In fact, cilia and flagella are specialized structures that can be found as protuberances in many bacteria (unicellular prokaryotic

---

[1] Cnidarians and ctenophores are two of the oldest living groups of animals and, although they are made up of multiple species, all of them have a gastrovascular cavity, often branched, consisting of the gastric cavity itself or central corridor and an orifice or opening that acts as both mouth and anus, often surrounded by tentacles. Examples of cnidarians are jellyfish and sea anemones.

organisms). These structures have receptors and ion channels on their surface that are capable of sensing changes in pH, temperature, or chemical elements and electrolytes in the environment. When activated, these receptors promote a structural change in the internal organization of the cilium or flagellum, triggering mechanisms to start or stop movement, thus affecting the direction taken by the bacterium during its locomotion. This event occurs automatically. The bacterium does not decide where to go, rather, environmental conditions cause the movement of the bacterium—similar to leaves moved by the wind, stones moved by the flow of a river, or the direction popcorn takes when it pops. So, in which animal can we start talking about pain?

The application of chemical (e.g., an acid), thermal (e.g., too high or too low temperatures), or mechanical (e.g., a pinprick or high pressure on a hand or foot) stimuli, that the vast majority of us would perceive as painful and try to avoid, can also lead to a change in behavior in certain animals. In other words, many animals exhibit aversive behaviors in response to identical stimuli that would cause pain in humans. It is by studying their behavior that we interpret what patients, in this case the animals, tell us.

It has been seen that a fly approaches and rubs its legs against its proboscis innumerable times in the course of a minute, but when some *wasabi* is placed on them, the number of approaches of the proboscis to the legs decreases abruptly. Similarly, it has also been found that if a number of flies are placed in an opaque tube with one end plunged in darkness and the other open to the light, they will instinctively tend to go towards the light. However, if all the flies on the dark side of the tube are gently tapped and a spot in the middle of the tube is heated with a light, no fly will dare to cross to the side open to the light. Another invertebrate, the octopus, hides its tentacle when it is hurt using tweezers or an electric shock by placing it far away from the aggressor source, and rubbing it repeatedly with one of its other tentacles. Diluted vinegar (acetic acid) was injected at the base of the antenna of shrimp, and in the nostrils of fish. In the former, it elicited a profuse grooming, or vigorous rubbing against the aquarium glass precisely over the injection site—the same behavior occurs if the antenna is pinched hard. In the latter, an increase in the frequency of opening/closing of the opercula per minute, and the repeated rubbing of its snout against the gravel at the bottom of the aquarium. Where is the fine line between simple perception, discomfort, and pain?

Precisely because our estimates of what an animal is suffering from are based solely on observations, we are often truly just comparing ourselves and empathizing. In many cases, then, we are dealing with assumptions. We compare and empathize all the more with those animals that are more similar to us. Most likely, nobody stops to think if a cockroach suffers, or feels pain.

Returning to the case of the crabs in the pot, studies on electrophysiology have failed to demonstrate the existence of thermoceptors (temperature receptors) in decapod crustaceans. So, for the moment, everything would point to the fact that crabs do not have neurons specialized in the transmission of information caused by thermal stimuli that would be harmful to humans. Does this mean then that they do not suffer when they are put in a pot of boiling water? In 2017, a video surfaced on social networks in which a crab about to fall into a boiling pot tore off its leg that was trapped by the lid of the pot with its contralateral claw to avoid falling into the pot. Could the crab perceive the excess temperature in some other way? Or is it an automated behavior in the face of danger, instinct, fear, or stress? Could it be that it was perceiving the distress of its companions? While this dramatic action has also been observed in different scenarios in the wild, the truth is that we do not have an answer for everything. We tend to think that those animals that present a structure more similar to us (in this case a nervous system) are also able to feel like us. However, fish breathe yet they do not have lungs.

According to the string of examples cited above, flies might simply dislike the taste of wasabi, find high temperatures unpleasant, or their instinct might tell them directly that heat of a certain intensity is not good for their viability, i.e., too much heat kills. It is well known that leeches have an oral suction cup with which they attach themselves to their prey, and through which they release mucus containing anticoagulants and anesthetics, preventing the prey from being aware of the leeches' presence. It is less well known that octopuses are capable of releasing mucus composed essentially of different glucans (sugars) and glycoproteins in combination with sialic acid or fucose (elements also present in human breast milk), which seems to have protective functions against microbial infections. That is, the octopus may release mucus and rub it on its injured tentacle to protect it from possible infection. Similarly, a shrimp or a fish might perceive a very intense stimulus on its antenna or on its snout that it knows it must get rid of, and may, consequently, try to rub its antenna or snout against the glass or gravel of its tank to get rid of the stimulus. In the case of the fish, the increased frequency of opening/closing of its opercula may be due to a situation of stress or nervousness, without necessarily implying suffering or pain. However, it is no less certain that all this *could* be pain. They cannot tell us. If it is already sometimes complex to know when something hurts a human, imagine how difficult it is to determine if something hurts an animal.

When we get a cramp in a muscle or a calf twitch, a relay race takes place in which a specialized neuron in direct contact with the affected muscle perceives the insult and converts (transduces) the mechanical stimulus into an electrical potential that is transmitted along its axon. Then, substances (neurotransmitters) are released at the other end of the neuron to stimulate the next nerve and thus continue passing the information to reach the goal: the brain—specifically, the *somatosensory cortex* (Fig. 1). This tells us the location, intensity, and duration of the stimulus (the cramp). However, the cortex is also reached by other biological signals coming from different areas of the brain. They stimulate anguish, fear, worry, alter memories, and relate the cramp to past experiences, state of mind, stress, tension, or depression. Through all these perceptions, at the end we have very detailed information of the importance that we should give to the perceived stimulus. It is the integration of all this information in the prefrontal cortex that makes up the subjective experience of pain, and all of this means that an injury is not perceived in the same way by each of us, or even in the same way by the same person at different times in their life. We also learn from pain. On the other hand, do not some people find pleasure in pain? The role of their primary specialized neurons (nociception or perception of the noxious stimulus) is the same: to perceive the intensity, location, and duration of the stimulus. However, the information coming from other different areas of the brain is, in a way, altered. In animals, the first part of the run is relatively easy to identify; however, the second part is much more difficult, especially depending on which animal. The only animal to which we can identify, appreciate, and quantify the participation of the different relays of pain transmission and perception is the human being, precisely because we can tell it to ourselves. "Perceiving" is not the same as "suffering."

Why would drugs that remove human pain allow flies to cross from the opaque side of the tube to the light side of the tube, regardless of whether the central part of the tube has been intensely heated? Would they prevent the shrimp (and the fish) from rubbing the area injected with acetic acid, and reduce the rate of opening/closing of the fish's opercula? Are we relieving pain in them or simply blocking the transmission/perception of the potentially noxious stimulus? Does it make sense to study only the first part of the race in such "underdeveloped" animals? The answer can possibly be found in dTRP (*Drosophila Transient Receptor Protein*). This is the gene originally identified in the fruit fly that codes for a protein that promotes/facilitates the transmission of harmful signals. Soon, a whole analogous gene family was discovered in other animals, and more importantly, drugs have been designed that act on some members of this family and prevent the transmission of specific nociceptive signals without affecting the perception of touch, pressure or motor response (movement). That is, from genetic and behavioral studies in flies, it has been possible to treat a specific type of pain in humans.[2] Whether or not flies perceive pain, they share similar (or homologous) actors and molecular mechanisms to those present in humans along our nerve transmission and perception pathways.

According to the annual death statistics report prepared by the National Statistics Institute of Spain (INE) when sorted by cause of death, heart disease and cancer are the leading causes of death. However, what the report does not include is that the main cause of doctor visits (along with other chronic diseases) is, precisely, pain. This is perhaps because one does not die of pain. One dies in pain, but not of pain. There are pains that, although initially simple, become difficult to treat or eradicate, especially if they become chronic. Therefore, the only way we can continue to advance our knowledge of the different types of pain is through

---

[2] The Qutenza®skin patch is a dressing impregnated with capsaicin 8%, the active substance in hot peppers. It binds to TRPV1 receptors, which are present exclusively in sensitive neurons, such as nociceptive neurons, which transmit noxious information. At first, capsaicin promotes an intense release of nociceptive substances (mainly substance P). However, this empties the nerve terminals of these nociceptive substances leaving the neuron silenced and unable to signal until no more substance P is synthesized. This produces a neuronal desensitization of these fibers, which no longer send information to the brain, causing the pain to cease.

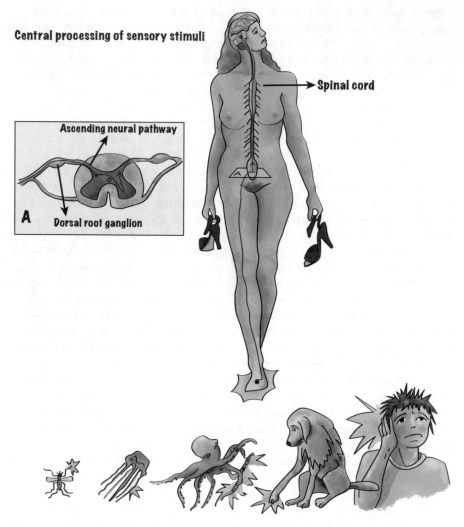

**Fig. 1 Conduction of sensory information**. Activation of nociceptive sensory fibers results in a chain of events that culminates in the perception of the stimulus in the brain. Skin nociceptors respond to acute stimuli that threaten to damage tissue integrity directly, through the transduction of stimulus energy by receptors located on nerve terminals or, indirectly, through the activation of channels and/or the release of molecules (alarmins) that, in turn, act on receptors located on the sensory neurons themselves. The stimulus, now encoded, travels along these primary neurons as nerve impulses and triggers the release of neurotransmitter vesicles at the opposite end of the nerve at the synapse with a secondary neuron in the dorsal horn of the gray matter of the spinal cord. The information then moves up through the sensory tracts and is transmitted to the thalamus, where a third neuron will finally stimulate specific areas of the somatosensory cortex producing the conscious perception of the initial stimulus. These mechanisms of transduction, transmission, and sensory perception are highly conserved throughout the evolution of the animal kingdom, from which it is inferred that the sensory system is essential for animal survival

studies in humans (these studies are, logically, limited because we cannot do certain things), or through the use of animal models of pain[3] in which nociceptive stimuli are induced. Changes in behavior or alterations produced at the molecular, cellular, histological, and physiological levels are then assessed, with their eventual comparison with humans. However, what happens on other sides of the die?

## The Platypus and Its Surprising Sexual Chromosome System

In nature, there are many different dice; that is, there is a great diversity of organisms thanks, in essence, to sexual reproduction, which allows the genetic information of the progenitors to be mixed, thus producing different offspring. The cells of our body are continually dividing (reproducing), making identical copies of themselves in a process called *mitosis*, by which we regenerate cells, and in turn tissues and thus organs. This happens constantly, for example, when we shed our skin or when we produce blood cells. The same type of reproduction occurs in other single-celled organisms, such as *protists*. This type of reproduction is asexual and occurs by mitosis. However, for eukaryotic organisms to reproduce, we use special cells called gametes. In sexual reproduction, two gametes, one from the father and one from the mother, fuse to create a new individual. The cells that will form the gametes divide by a special process called *meiosis*,[4] in which new genetic combinations are produced.

In eukaryotic organisms, genetic material is found as a disorganized ball protected inside a compartment, (the nucleus) as if it were a safe; when gametes are going to be formed, meiosis occurs in sexual or germinal cells. Following this analogy, the safe would open, that is, the nucleus would disappear momentarily, and this genetic material would condense, reorganizing itself into the well-known

---

[3] An animal model is used to study the functioning of a biological mechanism or a disease because, due to a series of specific characteristics, aspects of the pathology found in humans can be easily extrapolated. Thus, for example, pigs have a heart and a female reproductive system very similar to that of people, so that they can serve as training and thus prevent recent medical graduates from facing a cardiac catheterization or ovarian surgery without previous experience. However, it should be noted that, when trying to study the effect of a potentially useful drug for the treatment of a disease, a major difficulty is that molecules that are active in vitro (in cell cultures or in preparations with isolated tissues from animals) or in vivo, in the animal itself, have little chance of showing a similar absorption, therapeutic, and adverse effects profile in subsequent studies in humans. We are anatomically and physiologically different and, although many products end up working, a lot of time and resources are inevitably wasted studying other compounds that, in the end, will be useless to society.

[4] Meiosis is a specialized type of division whose aim is to produce gametes and create genetic variability in all populations of the planet; it is very important for evolution, and to preserve all those genetic modifications that improve a species and increase its chances of survival in this changing world. It consists of two consecutive divisions (without DNA replication in between), which reduces the genetic material of the cells by half and therefore we start from diploid cells (2n) and when producing gametes we obtain haploid cells (n).

chromosomes—very compact and dynamic little packages that contain all the information that makes it possible for our bodies to form. In the case of gametes, i.e., sex cells (sperm and eggs, to be more precise), when the packets of information are to be formed, they can be shuffled, as if they were a deck of cards. This process is called *recombination,* and a failure here could lead to alterations in the meiosis process, and thus to changes in the resulting gametes, or even their absence in the new generation of cells. The chromosomes then condense and arrange themselves in such a way that they can migrate to opposite poles of the cell, thus giving rise to two recombined daughter cells. This is followed by a second meiotic division (without DNA replication), generating four daughter cells with half the genetic content (n) and different from each other. Consequently, each egg and sperm (female and male gametes, respectively) are unique in their gene combinations, and this is how we each become different entities: when a particular gamete from the father and another from the mother maintain a special sex chromosomal relationship (Fig. 2).

Species have different numbers of chromosomes—both autosomes and sex chromosomes[5]—in the cells that make up their bodies, being reduced by half in sex cells, which instead of two have only one copy of each chromosome. Does the number of chromosomes have something to do with the complexity of the species at the evolutionary level? While humans have a total of 46 chromosomes (a total of 23 grouped in homologous pairs), mice have 40 chromosomes (20 pairs). However, apparently simpler organisms can have many more, such as shrimp, which have up to 90 chromosomes, or some butterflies, which have more than 200 chromosomes. Is it, therefore, the information contained in the chromosomes and not the number of chromosomes that truly makes us different? Looking exclusively at the sex chromosomes, in the case of women, there are two copies of the X chromosome, while in men, there is one copy of the X chromosome and one copy of the Y chromosome. X is larger and has more genes, while Y is smaller and has fewer genes. It is actually a degenerate copy of the X (sorry, guys!); however it is critical in making the individual who carries it male.

In nature, there are a multitude of very varied mutations that can affect all types of cells and chromosomes. They can affect autosomes, as occurs in Down syndrome, caused by the presence of an extra copy of chromosome 21 (or part of it), so it is also called trisomy of the 21st pair. It can also affect sex chromosomes, such as Klinefelter syndrome, when a male child is born with at least one extra X chromosome, XXY. Many of them have caused characteristics that have made the reproduction or survival of the species unfeasible, but others have forced the individual to acquire some characteristic configurations that have helped it to adapt to a changing environment.

---

[5] Sex chromosomes are chromosomes involved in determining the sex of the individual as male or female. The remaining chromosomes, called autosomal chromosomes, contain most of the genetic information of the individual, but are not critical in determining individual's sex.

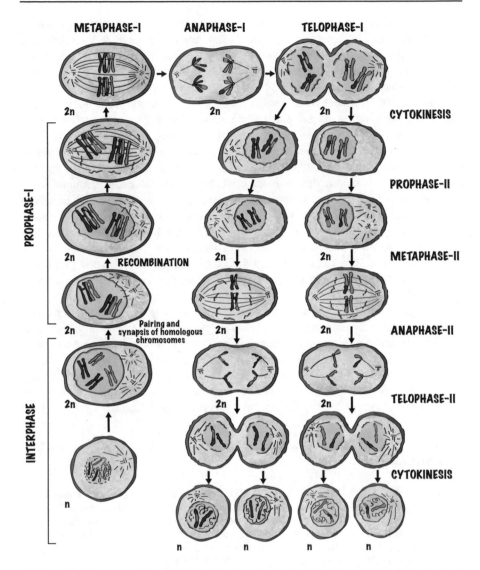

**Fig. 2  Sexual reproduction (meiosis) in the human body**. The different phases of meiosis are shown. Cells in the body have two sets of homologous chromosomes; they are diploid (2n) cells. In contrast, sex cells (gametes) have only one set of chromosomes; they are haploid (n)

As we have just seen, sex chromosomes are involved in sex determination, and knowing this is important because, for a large number of organisms, their sex determination depends on these chromosome systems and curious configurations. In grasshoppers, for example, males and females have different chromosomal endowments. Females are XX, while males have only one chromosome, being $X_0$.

In humans, the chromosome that determines male sex is Y; however, in grasshoppers, its absence defines sex. We have found rare cases beyond insects. Some species of kangaroos (swamp *wallabies*) also have odd sex chromosomes: males are $XY_1 Y_2$, and females are XX. This is due to the translocation of an autosomal pair (which previously had nothing to do with sex determination) to the X chromosome of the males of this *wallaby*. Despite this rarity, these adopted configurations are viable, and are transmitted generation after generation, allowing them to survive today. Curious, is not it? It does not stop there. Undoubtedly, the most extraordinary case is that of the platypus. This animal is already unique. It is semiaquatic, with a beak like a duck, a tail similar to that of a beaver and the legs of an otter. It is also the only mammal that produces venom, although only the males do. As if this were not enough, it is an oviparous mammal that both lays eggs and gives milk to its young, although it does not have breasts (the milk comes out of pores in its skin). To top it all off, it has been discovered that they have one of the most complex sex determination systems among animals, as they have a *cocktail* of 10 sex chromosomes: males 5X and 5Y and females 10X (as in parcheesi, sometimes everything starts with a 5, but you do not know how many turns the dice can take).

A simple chromosomal translocation can change the game. Thus, our mechanisms of sexual transmission and nociception, perhaps not in their entirety but at least in part, are conserved in all animals. Therefore, it is only by taking a humble view of ourselves as a species—comparing ourselves to other organisms—that we will be able to deal effectively with the evils that beset us. Although different, the essential mechanisms for life overlap, and present common patterns between species beyond chance. Know the past to understand the future.

## Bibliography

Accogli G, Scillitani G, Mentino D, Desantis S (2017) Characterization of the skin mucus in the common octopus Octopus vulgaris (Cuvier) reared paralarvae. Eur J Histochem 61(3):2518

Al-Anzi B, Tracey WD Jr, Benzer S (2006) Response of Drosophila to wasabi is mediated by painless, the fly homolog of mammalian TRPA1/ANKTM1. Curr Biol 16(10):1034–1040

Babcock DT, Landry C, Galko MJ (2009) Cytokine signaling mediates UV-induced nociceptive sensitization in Drosophila larvae. Curr Biol 19(10):799–806

Barr S, Elwood RW (2011) No evidence of morphine analgesia to noxious shock in the shore crab, Carcinus maenas. Behav Processes 86(3):340–344

Calvente A, Viera A, Parra MT, de la Fuente R, Suja JA, Page J, Santos JL, de la Vega CG, Barbero JL, Rufas JS (2013) Dynamics of cohesin subunits in grasshopper meiotic divisions. Chromosome 122:77–91

Correia AD, Cunha SR, Scholze M, Stevens ED (2011) A novel behavioral fish model of nociception for testing analgesics. Pharmaceuticals (Basel) 4(4):665–680

Cortez D, Marin R, Toledo-Flores D, Froidevaux L, Liechti A, Waters PD, Grützner F, Kaessmann H (2014) Origins and functional evolution of Y chromosomes across mammals. Nature 508:488

Crook RJ, Hanlon RT, Walters ET (2013) Squid have nociceptors that display widespread long-term sensitization and spontaneous activity after bodily injury. J Neurosci 33(24):10021–10026

Crow JF (1994) Advantages of sexual reproduction. Dev Genet 15(3):205–213

Daish T, Casey A, Grützner F (2009) Platypus chain reaction: directional and ordered meiotic pairing of the multiple sex chromosome chain in Ornithorhynchus anatinus. Reprod Fertil Dev 21(8):976–984

Della Rocca G, di Salvo A, Giannettoni G, Goldberg ME (2015) Pain and suffering in invertebrates: an insight on cephalopods. Am J Anim Vet Sci 10(2):77–84

Demin KA, Meshalkina DA, Kysil EV, Antonova KA, Volgin AD, Yakovlev OA, Alekseeva PA, Firuleva MM, Lakstygal AM, de Abreu MS, Barcellos LJG, Bao W, Friend AJ, Amstislavskaya TG, Rosemberg DB, Musienko PE, Song C, Kalueff AV (2018) Zebrafish models relevant to studying central opioid and endocannabinoid systems. Prog Neuropsychopharmacol Biol Psychiatry 86:301–312

Diggles BK (2019) Review of some scientific issues related to crustacean welfare. ICES J Mar Sci 76(1):66–81

Elwood RW, Appel M (2009) Pain experience in hermit crabs? Animal Behav 77(5):1243–1246

Enguita-Marruedo A, Martín-Ruiz M, García E, Gil-Fernández A, Parra MT, Viera A, Rufas JS, Page J (2019) Transition from a meiotic to a somatic-like DNA damage response during the pachytene stage in mouse meiosis. PLoS Genet 15(1):e1007439

Fernández-Donoso R, Berríos S, Rufas JS, Page J (2010) Marsupial sex chromosome behaviour during male meiosis. In: Deakin JA, Waters PD, Marshall Graves JA (eds) Marsupial genetics and genomics. Springer, Dordrecht, pp 187–206

Fiorito G, Affuso A, Anderson DB, Basil J, Bonnaud L, Botta G, Cole A, D'Angelo L, de Girolamo P, Dennison N, Dickel L, di Cosmo A, di Cristo C, Gestal C, Fonseca R, Grasso F, Kristiansen T, Kuba M, Maffucci F, Manciocco A, Mark FC, Melillo D, Osorio D, Palumbo A, Perkins K, Ponte G, Raspa M, Shashar N, Smith J, Smith D, Sykes A, Villanueva R, Tublitz N, Zullo L, Andrews P (2014) Cephalopods in neuroscience: regulations, research and the 3Rs. Invert Neurosci 14(1):13–36

Graves JAM (2018) Weird animals, sex, and genome evolution. Annu Rev Anim Biosci 6:1–22

Graz B, Diallo D, Willcox M, Giani S (2005) Screening of traditional herbal medicine: first, do a retrospective study, with correlation between diverse treatments used and reported patient outcome. J Ethnopharmacol 101(1–3):338–339

Grützner F, Rens W, Tsend-Ayush E, El-Mogharbel N, O'Brien PC, Jones RC, Ferguson-Smith MA, Marshall Graves JA (2004) In the platypus a meiotic chain of ten sex chromosomes shares genes with the bird Z and mammalian X chromosomes. Nature 432(7019):913–917

Hickman Jr CP (2006) Radiated animals. In: Hickman Jr CP, Roberts LS, Larson A, l'Anson H, Eisenhour DJ (eds) Comprehensive principles of zoology, 13th ed. McGraw-Hill/Interamericana de España, S.A.U., Madrid, pp 292–323

Hwang RY, Zhong L, Xu Y, Johnson T, Zhang F, Deisseroth K, Tracey WD (2007) Nociceptive neurons protect Drosophila larvae from parasitoid wasps. Curr Biol 17(24):2105–2116

Kang K, Pulver SR, Panzano VC, Chang EC, Griffith LC, Theobald DL, Garrity PA (2010) Analysis of Drosophila TRPA1 reveals an ancient origin for human chemical nociception. Nature 464(7288):597–600

Kim SE, Coste B, Chadha A, Cook B, Patapoutian A (2012) The role of Drosophila Piezo in mechanical nociception. Nature 483(7388):209–212

Lis-Kuberka J, Kątnik-Prastowska I, Berghausen-Mazur M, Orczyk-Pawiłowicz M (2015) Lectin-based analysis of fucosylated glycoproteins of human skim milk during 47 days of lactation. Glycoconj J 32(9):665–674

Magee B, Elwood RW (2016) No discrimination shock avoidance with sequential presentation of stimuli but shore crabs still reduce shock exposure. Biology Open 5:883–888

Manev H, Dimitrijevic N (2005) Fruit flies for anti-pain drug discovery. Life Sci 76(21):2403–2407

Mather JA (2016) An invertebrate perspective on pain. Animal Sentience 3(12) [online]. Available at: https://animalstudiesrepository.org/animsent/vol1/iss3/12/

Milinkeviciute G, Gentile C, Neely GG (2012) Drosophila as a tool for studying the conserved genetics of pain. Clin Genet 82(4):359–366

Nakagawa H, Hiura A (2006) Capsaicin, transient receptor potential (TRP) protein subfamilies and the particular relationship between capsaicin receptors and small primary sensory neurons. Anat Sci Int 81(3):135–155

Neely GG, Keene AC, Duchek P, Chang EC, Wang QP, Aksoy YA, Rosenzweig M, Costigan M, Woolf CJ, Garrity PA, Penninger JM (2011) TrpA1 regulates thermal nociception in *Drosophila*. PLoS ONE 6(8):e24343

Snow PJ, Plenderleith MB, Wright LL (1993) Quantitative study of primary sensory neurone populations of three species of elasmobranch fish. J Comp Neurol 334:97–103

Sneddon LU (2002) Anatomical and electrophysiological analysis of the trigeminal nerve in a teleost fish, Oncorhynchus mykiss. Neurosci Lett 319:167–171

Sneddon LU (2009) Pain perception in fish: indicators and endpoints. ILAR J 50(4):338–342

Snow PJ, Plenderleith MB, Wright LL (1993) Quantitative study of primary sensory neurone populations of three species of elasmobranch fish. J Comp Neurol 334(1):97–103

Stevens CW (2011) The evolution of vertebrate opioid receptors. Front Biosci 14:1247–1269

Tobin DM, Bargmann CI (2004) Invertebrate nociception: behaviors, neurons and molecules. J Neurobiol 61(1):161–174

Xu SY, Cang CL, Liu XF, Peng YQ, Ye YZ, Zhao ZQ, Guo AK (2006) Thermal nociception in adult *Drosophila*: behavioral characterization and the role of the painless gene. Genes Brain Behav 5(8):602–613

Zickler D, Kleckner N (2015) Recombination, pairing, and synapsis of homologues during meiosis. Cold Spring Harb Perspect Biol 7(6):a016626

**Miguel M. Garcia** holds a degree in Biochemistry from Universidad Autónoma de Madrid (UAM), a master's degree in the Study and Treatment of Pain, and a Ph.D. in Pain Research from Universidad Rey Juan Carlos (URJC). He is currently Assistant Professor and Researcher in the Area of Pharmacology, Nutrition and Bromatology of the Department of Basic Health Sciences at University Rey Juan Carlos. He belongs to the High Performance Research Group in Experimental Pharmacology (PHARMAKOM), and coordinates the Teaching Innovation Group in Diseases and their Treatment (EducaPath) of Universidad Rey Juan Carlos. He is a member of the International Association for the Study of Pain (IASP), the Spanish Pain Society (SED) and the Spanish Society of Pharmacology (SEF). His research work focuses on basic pharmacology in the field of pain, particularly on the role of glial cells and cannabinoid and TLR4 receptors in nociception. As a disseminator, he has experience as a speaker at different science festivals and activities.

**Marta Martín Ruiz** has a degree in biology from Universidad Complutense de Madrid (UCM), a master's degree in Pain Study and Treatment from Universidad Rey Juan Carlos (URJC), and a Ph.D. in Biology from Universidad Autónoma de Madrid (UAM). After having conducted a postdoctoral internship at the Magee-Womens Research Institute & Foundation located in Pittsburgh, USA, where she focused on the regulation of gametogenesis in humans and mice and, more specifically, on the fundamental mechanisms required to produce viable germ cells, she currently leads the Laboratory of Molecular Biology at Hospital Universitario General de Villalba in Collado Villalba, Madrid. She has taken part in open science and dissemination events such as *Pint of Science*, *Science Week*, *World Brain Week* and *Science Marathon*.

# Flies Do Not Like Wasabi

Miguel Molina Álvarez⊙
and Blanca del Carmen Migueláñez Medrán⊙

Throughout evolution, different organisms have developed a series of strategies to defend themselves from the unwanted attack of their predators; strategies that evolved continuously and in parallel to the new abilities of the predators, establishing a kind of "cat-and-mouse chase" over time. In some cases, defense strategies do not develop fast enough, and the cat traps the mouse. When it comes to real life, it is not that a plant tries to outrun pests or predators; however, over generations, genetic mutations occur which sometimes, as a matter of chance, protect the individuals with that mutation against the predator, providing a greater chance of survival against their peers (*evolutionary advantage*)—until a predator develops new skills to deal with it. Thus, by the vagaries of nature, certain plants have managed to design various chemical agents that produce predator effects such as stinging, irritating, burning or painful sensations. However, how are plants able to produce these effects on organisms?

M. Molina Álvarez (✉)
Area of Pharmacology, Nutrition and Bromatology, Department of Basic Health Sciences, Unidad Asociada de I+D+i al Instituto de Química Médica (IQM-CSIC), Universidad Rey Juan Carlos, Madrid, Spain
e-mail: miguel.molina@urjc.es

High Performance Experimental Pharmacology Research Group (PHARMAKOM), Universidad Rey Juan Carlos, Alcorcón, Spain

Grupo Multidisciplinar de Investigación y Tratamiento del Dolor (i+DOL), Alcorcón, Spain

B. C. Migueláñez Medrán
Area of Stomatology, Department of Nursing and Stomatology, Universidad Rey Juan Carlos, Alcorcón, Spain
e-mail: blancac.miguelanez@urjc.es

© The Author(s), under exclusive license to Springer Nature Switzerland AG 2024     137
M. M. Garcia (ed.), *Tales of Discovery*, https://doi.org/10.1007/978-3-031-47620-4_11

Not only plants, but predators (including humans), also possess different defense mechanisms. One example is taste chemoreceptors, which is a protective measure against the ingestion of possible "food" that could constitute a health risk; another example is the portal-hepatic system, by which any substance—or a considerable fraction of it—introduced into our digestive tract is returned to the liver so that it may be metabolized (deactivated) before entering the circulatory system. It is therefore logical to think that if the ingestion of a plant causes pain or a burning sensation, we will stop eating it immediately. We interpret it as an alarm signal. The curious and fascinating thing is that plants that are not toxic to the body produce unpleasant sensations when they come into contact with sensitive neurons in the mouth; this is actually a strategy developed by the plant to try to deceive the predator, and thus avoid being eaten. These plants, such as hot peppers (chili) are very popular in the culinary world. Although the sensations they produce have been long described, for years *why* they were generated was an enigma. To understand this, we have to move down to the cellular and molecular levels. Only then will we understand how plants manage to produce these sensations.

## TRP or Transient Receptor Potential Ion Channels

It was not until 1969 that researchers Cosens and Manning discovered TRP receptors in the fruit fly (*Drosophila melanogaster*). These are a family of receptors, located in neurons responsible for transmitting noxious (or potentially dangerous to the body) sensations, that functions by activating gates that open and allow specific positively charged ions (*cations*) to pass through them. The main and most studied cation is calcium,[1] which enters the cellular interior by way of a concentration gradient. The entry of cations generates a change of polarity on both sides of the cell membrane and causes the information to travel to the brain, producing one sensation or another depending on which TRP receptor has been activated, and in which neuron (heat, cold, itch, pain, etc.).

How does this happen? Basically, the inside and outside of the cell as a whole are essentially neutral, with electrolytes (free ions) generally bound to charged protein or lipid molecules (organic ions), and the charges counteract each other. However, the inner side immediately in contact with the membrane has a small number of negative charges higher than that of the outer side, while the outer side has a small number of positive charges higher than that found on the inner side. In this way, the membrane behaves like a capacitor—a structure that keeps negative and positive charges separated and prevents them from coming together. This gives rise to an electric potential difference—a voltage. Therefore, under basal, resting

---

[1] Within the extensive TRP receptor superfamily (TRPA1-TRPA2, TRPC1-TRPC7, TRPM1-TRPM8, TRPML1-TRPML3, TRPP1-TRPP3, TRPV1-TRPV6), each member has different permeability to different cations, including $Ca^{2+}$, $Fe^{2+}$, $Mg^{2+}$, $Sr^{2+}$, $Ba^{2+}$, $Mn^{2+}$, $Zn^{2+}$, $Ni^{2+}$, $Co^{2+}$, $Cd^{2+}$, $La^{3+}$, $Gd^{3+}$ or $Fe^{3+}$. For a more extensive review see: Bouron et al. Pflugers Arch. 2015;467(6):1143–64.

conditions, every membrane has a certain voltage; each cell a source of potential energy, and the difference is maintained because the membrane is selectively permeable.

In the membrane of a cell, there are, on the one hand, a number of channels that are permanently open, such as certain potassium channels. Thus, one would tend to think that potassium is continuously leaving the cell—that the cell loses potassium. However, the cell does not empty itself of potassium because organic ions, which cannot cross the membrane, exert an attractive force on the free potassium ions. On the other hand, there are also a series of channels crossing the cell membranes whose opening is conditioned on the intervention of certain stimuli. Examples include: *ligand-regulated channels*, in which a molecule binds to the cell membrane causing a conformational change, and therefore its opening; *voltage-regulated channels*, in which changes in the membrane potential (modifications in the voltage difference between both sides of the membrane) also induce a conformational change, and lead to their opening; or, *mechanosensitive channels* that open in response to a mechanical stimulus.

There are different types of voltage-regulated channels, classified according to their majority permeability to a certain type of ion (sodium, potassium, calcium, chlorine, etc.). Each of these ions is present in a notably different concentration on the outside and inside of the cell. Thus, for example, the concentration of sodium in the extracellular medium (145 mM) is approximately ten times higher than that in the interior (15 mM), while that of potassium is almost forty times higher in the intracellular medium (150 mM) than in the extracellular medium (4 mM). Therefore, as would happen in a full subway car, in order for more passengers to enter, some passengers must first exit. In our context, it means that there is a concentration gradient *from* the compartment with the highest concentration *to* the one with the lowest concentration: sodium will tend to enter the cell, and potassium will leave. The concentrations of sodium, potassium and other ions in the cells of different animals may vary, but the different ratios have been maintained throughout evolution; otherwise, the existence of a resting membrane potential (voltage difference) would be impossible.

When a voltage-regulated sodium channel opens, it creates a flow of sodium ions that migrate from the side where their concentration is higher (cell exterior) to the side where their concentration is lower (cell interior). With this action, positive charges disappear on the outer side immediately on contact with the cell membrane, and are gained on the inner side. That is, the polarity is reversed. However, this happens only for a very short fraction of time, only 2 ms, and in a specific area of the membrane. Voltage channels do not discriminate when a voltage difference occurs, so voltage-regulated potassium channels will also open, creating a flow of potassium ions that migrate from where their concentration is higher (inside the cell) to where their concentration is lower (outside the cell). This outflow of potassium ions, together with the outflow that occurs constitutively through potassium leak channels, allows a return to basal conditions with no net loss of cations (positively charged ions) (Fig. 1). Concomitantly, proteins called Na+/K+pumps, located in the cell membrane, re-establish the ratio of $Na^+$ and $K^+$ ions inside the

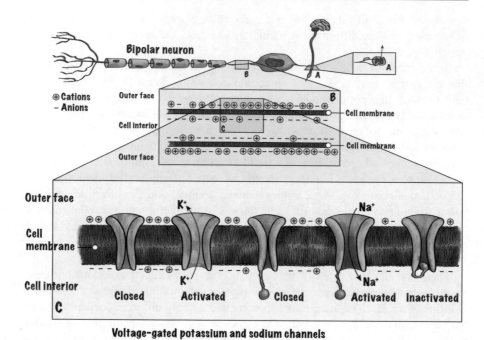

**Fig. 1** **The importance of voltage-gated channels in the creation of a membrane potential difference for proper transmission of information**. Throughout the various stages of the creation of an action potential, a polarity reversal is created at a particular point on the membrane, triggering the opening of the next voltage-regulated channel as the voltage of its immediate environment changes

cell by introducing two potassium ions for every three sodium ions pumped out. Although we have mentioned only sodium and potassium, all the electrochemical ion balances contribute to the resting membrane potential difference.

Within these *ion channels,* we find that the TRPs and, as mentioned above, their opening, would be modulated by a series of stimuli, which could be changes in pH, molecules (e.g., allicin, menthol, capsaicin or mustard oil), mechanical stimuli, or hot or cold temperatures, depending on the receptor subtype. This rapid entry of cations into the neuron would ultimately result in the transmission of a signal that would travel to the brain; after complex processing, the signal would be transformed into a different sensation, depending on the receptor and the neuron that had been activated in the periphery. In this way, it is possible that a chemical substance can activate a TRP designed to detect stimuli that are excessively hot (potentially harmful to the body) or cold, and cause us to feel as if we are burning, when in fact that is not the case (Fig. 2).

The TRP family is the most important and abundant group of receptors in the nociceptive fibers of different animals. In recent years, several studies have shown that the molecular pathways, by which the transduction of these signals that activate ascending pain pathways occurs, are highly conserved in animals

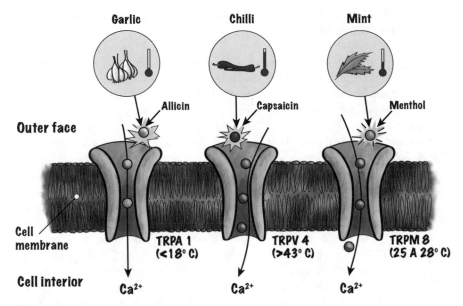

**Fig. 2** **Mechanisms of activation and regulation of TRP channels in sensory neurons**. TRPA, TRPV and TRPM channels present diverse mechanisms of activation. Regardless of the mechanism leading to their activation, TRPV1, TRPV2, TRPV3 and TRPV4 are able to emulate high temperature sensations. TRPA1 and TRPM8, on the other hand, induce intense cold sensations. In some cases, there is a presence of heterologous channels formed by subunits of different channels (e.g., TRPV1/TRPA1, TRPV4/TRPA1 or TRPV1/TRPV4), and there is a relationship in their regulation. The coexpression of TRPV1 and TRPA1 channels in the same cell, and the existence of heterologous channels leads to a cold or intense heat stimulus being perceived as a burning sensation. In addition, the activation and opening of these channels set in motion a complex intracellular signaling and regulation mechanism in which protein kinases (PKA, PKC, CAMKII, PI3K or Src PTK) act as secondary activators of these channels from the intracellular side. Although the figure shows the different types of channels in the same membrane, this does not mean that the different TRPs are all expressed in the same cell

ranging from insects to humans. To date, six different types of TRPs have been discovered in the plasma membranes of mammalian peripheral sensory neurons: TRPCs, TRPM, TRPV, TRPA, TRPP, and TRPML. Although there is a seventh type present in nonmammalian animals: TRPN. These receptors are very complex, and participate not only in the transmission of noxious stimuli of mechanical, chemical, or thermal origin, but are also found in most tissues and cell types. To date, they have been found to be involved in processes such as pain initiation, thermoregulation, salivary fluid secretion, inflammation, cardiovascular regulation, smooth muscle tone, pressure regulation, calcium and magnesium homeostasis, and lysosomal function. In addition, mutations in TRPs appear to be associated with abnormal pain sensitivities, gastrointestinal diseases, kidney disease, neurodegenerative diseases, certain types of cancer, asthma, and even psychiatric diseases. All of this makes TRP receptors unique. Although so much information may seem overwhelming, in this chapter we will learn more about these receptors.

## Chili, Garlic, Wasabi, and Menthol: The "Nouvelle Cuisine" in the Treatment of Pain

Regarding the transmission of nociception (potentially dangerous stimuli for the organism, which after being processed in the brain can produce the sensation of pain), there are three key TRP receptors: TRPV1, TRPA1 and TRPM8. These receptors are present in different types of primary sensory neurons, which are responsible for detecting stimuli, both external and internal, and for transferring the information to the spinal cord. These neurons can be differentiated according to the TRP receptors present in their membranes and, consequently, according to the type of stimuli that can lead to their activation. Thus, for example, the TRPV1 receptor is mainly expressed in sensory neurons with small and medium caliber axons. What does this mean? Sensory neurons with smaller caliber axons are the fibers responsible for transmitting noxious stimuli to the central nervous system. These are called C (small caliber) and A∂ (medium caliber). Therefore, those neurons that present TRPV1 receptors in their membrane will transmit nociception. Other receptors, such as TRPM8 and TRPA1, are also present in nociceptive neurons.

The presence of a certain receptor in a neuron can mean the absence of another receptor. For example, if a neuron has TRPM8 receptors on its membrane, it will not have TRPV1. However, it is common for TRPA1 and TRPV1 receptors to be expressed in the same cell. In fact, in the dorsal root and trigeminal ganglia, TRPA1 expression is practically restricted to sensory neurons that coexpress TRPV1. This could explain why a stimulus of intense cold that would activate TRPA1 receptors, or intense heat that would activate TRPV1 receptors, can both be perceived as a burning sensation. Alternatively, substances such as allicin, diallyl disulfide (DADS), or allyl isothiocyanate (AITC), organosulfur compounds present in garlic, wasabi, mustard, or radishes produce a burning sensation similar to the capsaicin present in hot peppers, which are capable of activating TRPV1 receptors. Therefore, determining which receptors are present in different populations of neurons can be of vital importance when searching for drugs that act in certain areas of the body, or that alleviate a particular sensation. However, the feature that makes these receptors especially attractive to researchers is that their continued activation over time produces a desensitization of the neurons that have this receptor, and sparks great interest as a possible treatment for pain. Have you ever perceived a very strong smell as soon as you enter a place, but after a while that small seems to have decreased in intensity? Or have you ever wondered why some people or cultures tolerate spicy food better than others? The answer seems to be that continuous exposure to a certain spicy food (which activates a TRP receptor) produces long-term desensitization in the population of neurons stimulated in the mouth.

The TRPV1 receptor is the most studied and, currently, of great interests in the treatment of pain. These channels are activated by temperatures above 43 °C and by various chemical agents, such as capsaicin (the active component of hot peppers) and other vanilloid compounds. Hence, the sensation of "fire" that we feel when we ingest a chili-derived spice, or when rubbing our eyes after touching a cayenne pepper. Capsaicin causes a painful sensation translatable into hyperalgesia to heat due to increased discharges in nociceptors C and Aδ, which express TRPV1. When capsaicin binds to the TRPV1 receptor of a neuron, it causes a series of changes that lead to the transduction of this signal to the brain, and produces the release of proinflammatory substances at the medullary level. Neurons that have TRPV1 in their plasma membrane react to this substance as if it were an aggression, not only transmitting information to the brain but also releasing substances that will produce neurogenic inflammation (inflammation of the nerve), which results in more signals being sent to the brain and, ultimately, more pain. These proinflammatory substances include glutamate, substance P, and calcitonin gene-related peptide (CGRP). It may be difficult to believe that a substance that, in the short term, produces an intensification of pain can serve as a treatment for pain. However, the massive opening of TRPV1 channels by capsaicin when consumed frequently produces a series of secondary consequences in the medium- and long-term in neurons. In the medium-term, proinflammatory substances that are stored in vesicles in the synaptic terminal are depleted, which renders the neuron insensitive and unable to transmit the signal to the next neuron even when stimulated. In the long-term, the continuous activation and opening of TRPV1 channels produces toxification due to excess calcium, rendering the neuron useless for a period of approximately six weeks.

Therefore, let us imagine that we have a constant pain in our leg and we apply a spicy substance on the skin. At first, we would feel a more intense pain than usual, but after a while this intense pain would disappear and with it, the initial pain. This is the principle on which capsaicin patches are based. However, it is not that simple, and this intentional desensitization does not work for every type of pain. The application of these patches is becoming increasingly common, but only for the treatment of neuropathic pain, where a nerve is affected (damaged or inflamed.

The TRPA1 receptor can be activated by temperatures below 18 °C, allicin, and allyl isothiocyanate (the former an active component in garlic; the latter an active component in mustard, horseradish, and wasabi). As in the case of TRPV1 receptors, their activation leads to the entry of cations into the cell interior, resulting in the release of proinflammatory substances. The fact that this channel is linked to the detection of noxious cold, even though the chemical agents mentioned do not cause cold sensations but rather burning sensations, could be explained precisely because many of the neurons that express TRPA1 also express TRPV1, and the resulting sensation is similar to that found after TRPV1 activation. It is for this reason that the sensation produced by garlic and wasabi is not the same as that produced by chili, but they all generate a burning sensation.

In recent years, several phylogenetic studies have been carried out to determine the evolutionary relationships between different groups of genes in different organisms. This is of great human interest mainly for two reasons. First, garlic, mint, or wasabi have been used by different cultures as natural repellents for flies and other insects. Isolating the substances responsible for this aversion would be economically advantageous for the food and pharmaceutical industries, especially since those substances lack relevant toxicity to humans. Second, it has been demonstrated that TRPA1 receptors act as chemical taste sensors to inhibit the ingestion of certain compounds in flies. In a study that has been repeated in recent years, flies were immobilized, and a drop of a solution containing allyl isothiocyanate (AITC), a chemical compound present in wasabi, applied to their front legs. The test consists of comparing the number of times the proboscis approaches the substance compared to a control group to which a neutral or sweetened solution was applied. It has been determined that the decrease in the number of proboscis approaches to the wasabi extract solution is not due to an olfactory or gustatory issue, as might be inferred, but is related to an avoidance behaviour to protect the fly against the ingestion of potentially toxic substances.[2]

In evolution, things tend to look less and less alike. Therefore, the development of assays in animal models may be helpful in discovering the role of genes and proteins complementary to TRPA1 in *Drosophila*. These will have diverged greatly in humans; however, they will help to explain the ability of some proteins to regulate genes that are distant from each other in a genome sequence, and may serve as new therapeutic targets. Interestingly, a rare inherited mutation of the TRPA1 channel has recently been linked to a disorder known as *familial episodic pain syndrome*. This mutation in the channel produces a gain of function that makes it more sensitive to activation and consequently produces intense pain under conditions of physical stress (such as prolonged fasting or fatigue), and exposure to cold. Additionally, the syndrome is accompanied by shortness of breath, tachycardia, sweating, pallor, and rigidity of the abdominal wall, which gives us an idea of how important these channels are for the proper functioning of the body.

Finally, menthol, an active compound of another culinary supplement, such as peppermint, can cause a sensation of hypersensitivity to cold when administered topically, due to the stimulation of small nociceptive fibers (C). In fact, both menthol and temperatures between 23 °C and 28 °C are able to activate TRPM8 channels, inducing neuronal activation. In pathologies such as diabetes and cancer, in which patients may present hypersensitivity to cold, it seems that the TRPM8 channel plays a fundamental role. The presence of menthol is widespread in many anti-inflammatory and analgesic creams, since the sensation of intense cold produces a desensitization of the area being treated, enhancing the action of the drug and helping to improve the patient's symptomatology.

As we have just observed, what sets the therapeutic strategy for these channels apart is that it is no longer directed at the laborious and costly design of new drugs intended to block these receptors; rather, the algesic agent itself can be used to desensitize these receptors and consequently inhibit nociceptive activity. Because

---

[2] The experiment involves fasting a series of flies overnight, after which they are anesthetized and immobilized on a surface. The flies are then hydrated and a drop of a solution is placed on the front legs (e.g., sucrose or sucrose + AITC). A scoring system is created whereby, if the proboscis extends and contacts the drop for 2–3 s, a value of 1 is awarded; if contact is brief or simply sniffing occurs, 0.5 points are awarded; if there is no contact within 5 s, 0 is assigned. The final score is the result of five presentations 2 min apart.

the ingredient that activates the channel is the same as the one that desensitizes it, costs are very low, making these receptors very interesting targets in the treatment of neuropathic pain. Although the popularity of these receptors is relatively new, the *Hungarian School* has been using pharmacological experiments for more than 60 years that show the paradoxical analgesic role of capsaicin; however, the lack of knowledge of cellular and molecular techniques prevented them from carrying out more advanced studies.

# Bibliography

AN Akopian NB Ruparel NA Jeske KM Hargreaves 2007 Transient receptor potential TRPA1 channel desensitization in sensory neurons is agonist dependent and regulated by TRPV1-directed internalization J Physiol 583 Pt 1 175 193

B Al-Anzi WD Tracey S Benzer 2006 Response of *Drosophila* to wasabi is mediated by painless, the fly homolog of mammalian TRPA1/ANKTM1 Curr Biol 16 10 1034 1040

DM Bautista P Movahed A Hinman HE Axelsson O Sterner ED Högestätt D Julius S-E Jordt PM Zygmunt 2005 Pungent products from garlic activate the sensory ion channel TRPA1 PNAS 102 34 12248 12252

A Bouron K Kiselyov J Oberwinkler 2015 Permeation, regulation and control of expression of TRP channels by trace metal ions Pflugers Arch 467 6 1143 1164

González-Ramírez R, Chen Y, Liedtke WB, Morales-Lázaro SL (2019) TRP channels and pain. In: Neurobiology of TRP channels. CRC Press, pp 125–148

Gould SJ (2010) The adaptationist program. The structure of evolutionary theory, 3rd ed. Tusquests Editores, Barcelona, pp 181–186

N Jancsó A Jancsó-Gábor J Szolcsányi 1967 Direct evidence for neurogenic inflammation and its prevention by denervation and by pretreatment with capsaicin Br J Pharmacol Chemother 31 1 138 151

K Kang SR Pulver VC Panzano EC Chang LC Griffith DL Theobald PA Garrity 2010 Analysis of *Drosophila* TRPA1 reveals an ancient origin for human chemical nociception Nature 464 7288 597 600

Levine JD, Alessandri-Haber (2007) TRP channels: targets for the relief of pain. Biochim Biophys Acta 1772(8):989–1003

SJ Mandel ML Shoaf JT Braco WL Silver EC Johnson 2018 Behavioral aversion to AITC requires both painless and dTRPA1 in *Drosophila* Front Neural Circuits 12 45

B Minke 2010 The history of the *Drosophila* TRP channel: the birth of a new channel superfamily J Neurogenet 24 4 216 233

M Numazaki M Tominaga 2004 Nociception and TRP channels Curr Drug Targets CNS Neurol Disord 3 6 479 485

LS Premkumar 2014 Transient receptor potential channels as targets for phytochemicals ACS Chem neurosci 5 1117 1130

K Roberts R Shenoy P Anand 2011 A novel human volunteer pain model using contact heat evoked potentials (CHEP) following topical skin application of transient receptor potential agonists capsaicin, menthol and cinnamaldehyde J Clin Neurosci 18 7 926 932

NB Ruparel AM Patwardhan AN Akopian KM Hargreaves 2008 Homologous and heterologous desensitization of capsaicin and mustard oil responses utilize different cellular pathways in nociceptors Pain 135 3 271 279

SA Springer BJ Crespi WJ Swanson 2011 Beyond the phenotypic gambit: molecular behavioural ecology and the evolution of genetic architecture Mol Ecol 20 11 2240 2257

J Szolcsányi 2004 Forty years in capsaicin research for sensory pharmacology and physiology Neuropeptides 38 6 377 384

K Venkatachalam C Montell 2007 TRP channels Annu Rev Biochem 76 1 387 417

L Vyklický K Nováková-Tousová J Benedikt A Samad F Touska V Vlachová 2008 Calcium-dependent desensitization of vanilloid receptor TRPV1: a mechanism possibly involved in analgesia induced by topical application of capsaicin Physiol Res 57 Suppl 3 S59 68

**Miguel Molina Álvarez** has a degree in Physical Therapy from Universidad Rey Juan Carlos (URJC), and a double master's degree in Manual Physiotherapy of the locomotor system from Universidad de Alcalá de Henares (UAH) and in *Clinical and Basic Aspects of Pain* from Universidad de Cantabria (UC)-Universidad Rey Juan Carlos. He currently combines his research work in animal models for the study and treatment of pain in the Area of Pharmacology, Nutrition and Bromatology of Universidad Rey Juan Carlos with his activity as a physiotherapist. He belongs to the High Performance Research Group in Experimental Pharmacology (PHARMAKOM) of Universidad Rey Juan Carlos and is a member of the Spanish Pain Society (SED).

**Blanca del Carmen Migueláñez Medrán** has a degree in dentistry, a master's degree in implant dentistry, and a Ph.D. in dentistry from Universidad Rey Juan Carlos (URJC). She is a member of the Spanish Society of Oral Medicine (SEMO). She currently combines her research and teaching work in Preventive and Community Dentistry (Degree in Dentistry) and in Oral Medicine (Master's Degree in Oral Medicine) with her activity as a dentist, for which she has been accumulating years of experience in private clinics over the past years.

# The Glia: Squires of the Nervous System

Miguel M. Garcia⊙ and Marina Martín Taboada⊙

Sancho Panza in *Don Quixote*, Iñigo Montoya in *The Princess Bride,* or Robin in *Batman*—each the slightly smaller role—a supporting character, but worthy of an Oscar. Each remains in the background, but sometimes steals the limelight from the main character.

The nervous system is made up of neurons (the main character) and of glial cells (the supporting characters). But what are glial cells? The term glia or neuroglia means something like glue,[1] and was first coined in 1858 by the German physician

---

[1] Actually, the term *glia* comes from the Latin *glūt* and this in turn from the Greek *gloiós* and is attributed to Semonides who made use of it to define the oily sediment for baths, viscous, or chewy texture. In English, *glue* derives from this term. and we still keep it in Spanish in the words *gluten* and *aglutinar*.

---

M. M. Garcia
Area of Pharmacology, Nutrition and Bromatology, Department of Basic Health Sciences, Unidad Asociada de I+D+i al Instituto de Química Médica (IQM-CSIC), Universidad Rey Juan Carlos, Madrid, Spain
e-mail: miguelangel.garcia@urjc.es

High Performance Experimental Pharmacology Research Group (PHARMAKOM), Universidad Rey Juan Carlos, Alcorcón, Spain

Grupo Multidisciplinar de Investigación y Tratamiento del Dolor (i+DOL), Alcorcón, Spain

M. Martín Taboada (✉)
Area of Biochemistry and Molecular Biology, Department of Basic Health Sciences, Universidad Rey Juan Carlos, Alcorcón, Spain
e-mail: marina.martin@urjc.es

High Performance Research Group in the Study of Molecular Mechanisms of Glucolipotoxicity and Insulin Resistance: Implications in Obesity, Diabetes and Metabolic Syndrome (LIPOBETA), Universidad Rey Juan Carlos, Alcorcón, Spain

Rudolf Virchow, who in his dissertation on *Cellular pathology based on the physiological and pathological study of tissues* during the 20th Congress of the Institute of Pathology in Berlin, referred to a mass of tissue, traditionally considered simply connective tissue, as *neuroglia* or *nervenkitt*—that is, neural cement. Currently, authors send a handwritten summary of their work to the organization hosting the congress they attend; that summary can be consulted in a book of abstracts almost as soon as the congress is held. In Virchow's time, however, the books of communications were prepared from the shorthand of the papers, and the publication of the book of proceedings of the congress took a few months to emerge. It was, therefore, not until months later that Virchow's neuroglia began to be appreciated as something different from the connective tissue, not just as a mere structural element of the tissue, and began to gain interest among the scientific community. The development of the modern microscope industry played a decisive role in the discovery and understanding of glia. Emerging physicists who increased the limit of optical resolution (e.g., Ernst Abbe), and chemists who improved the quality of the glass present in objectives and lenses (e.g., Otto Schott), gave rise to newly created companies such as the German company Carl Zeiss, (founded in 1846), making it possible to systematize the construction of microscopes. Since then, glia have been found in a great diversity of living beings, with varying degrees of diversification: from nematodes to humans, leeches, flies, and other animals. However, glial cells do not seem to be present in all animals (or at least, it has not yet been detected in some). An example of this is rotifers.

## Discovering the Glia

Human life is finite, so, despite individual awards and recognition, scientific progress is a collective task that results from the contribution of new discoveries to those already initiated by predecessors. As John of Salisbury wrote in his work *Metalogicon* more than 800 years ago, "we are like dwarfs standing on the shoulders of giants." Only by relying on the works and ideas of others can we glimpse new advances. In this respect, many of the milestones in the field of neuroglia have been the result of improvements in microscopy and the use of new staining techniques contributed by scientists from different disciplines, especially Germanic, between the second quarter of the nineteenth century and the twentieth century. It could be said that during these hundred years, *a school was created*. Thus, for example, in 1838 and 1839, Robert Remak and Theodor Schwann, with their description of myelin bands and the discovery of the cells that produce myelin, respectively, laid the foundations for the discovery of Schwann cells sixty years later (Louis-Antoine Ranvier, 1871). However, it took more than ten years to identify a glial cell. In 1851, Heinrich Müller, through his studies of comparative anatomy of the retina of various vertebrate species, described for the first time a type of large retinal glial cell that later became known as Müller cells (Max

Schultze, 1859). In the following years, descriptive works about non-neuronal cells with a stellate aspect (Otto Deiters, 1860; Jakob Henle, 1869) would abound, and two different types were noted: fibrous glia and protoplasmic glia (Albert von Kölliker, 1889; William Lloyd Andriezen, 1893). However, it would not be until the end of the nineteenth century when this type of cell would be classified as a new group of glial cells: astrocytes (Michael von Lenhossék, 1893).

Throughout the nineteenth century, the most widely accepted idea was that glia were passive, a mere structural support that filled in the spaces between neurons. This was embodied by Karl Weigert in his *infill theory* in 1895. However, precisely in the last years of that century, two new ideas began to gain strength: Golgi's *reticular doctrine* and Cajal's *neuronal doctrine*. The first proposed that the nervous system was made up of a network of fused cells in the form of a syncytium. This idea was supported by the work of other researchers, such as Carl Ludwig Schleich, who in the 1890s proposed that glia and neurons were both active elements arranged in interconnected circuits. Ten years later, in 1904, Hans Held postulated a syncytial theory, according to which glial cells formed a syncytial network in which they are all connected to each other. Golgi further hypothesized that glial cells, by virtue of their proximity to blood vessels and neurons, should function by carrying nutrients to them. For its part, the *neuronal doctrine* proposed that neurons were individual brain cells, and it marked the beginning of neuroscience. This theory also had numerous supporters. Wilhelm Gottfried Waldeyer and Sigmund Exner contributed to the consolidation of the *neuronal doctrine*, the former proposing in 1891 the name *neuron* for the cells described by Cajal. Cajal also devoted himself to the study of glial functions, and supported his brother Pedro's theory of isolation, in which astrocytes[2] would act as physical insulators of neurons, preventing the activity of one neuron from affecting others that it should not, thus putting a stop to undesirable signaling.[3]

The development of specific stains for neuronal and neuroglial cells, such as metal impregnation techniques, enabled histological differentiation between astrocytes, microglia, and oligodendroglia, helping to postulate distinct glial functions.

---

[2] Astrocytes or astroglia are the most numerous cells in the brain. Today, it is known that they control the birth, development, functional activity, and death of neuronal circuits—mother elements from which neurons are born. They integrate neurons, synapses, and capillaries into interdependent units. They constitute the internet, the mycelium, the mesh, the syncytium that functions as a unit—just like the connection between the trees in *Avatar*. They have been considered secondary cells, but they are essential elements.

[3] In 1906, Santiago Ramón y Cajal and Camillo Golgi received the Nobel Prize in Medicine and Physiology for their work on the structure of the nervous system and their contributions to the field of neuroscience. At that time, there were no cameras attached to microscopes as we have today, so researchers of the time had to not only be good physiologists or anatomopathologists, but also good illustrators. It is said that Cajal was not as good a draughtsman as Golgi, however, he was able to interpret what he thought was happening. By using various staining techniques, Cajal was able to virtually describe every part of the central nervous system.

Today, it is recognized that there are two main types of cells in the nervous system: neurons (electrically excitable) and glial cells (chemically excitable). Within both, several types have been described. In the case of glial cells, although they are all grouped under the same name, they have often been classified according to their origin: neural (macroglia: astrocytes, oligodendrocytes, and ependymal cells) or immune (microglia, Schwann cells, satellite cells, and glial cells of the enteric nervous system). Thus, all neural cells (neurons and macroglia) derived from the neuroepithelium that forms the neural tube, and their progeny in the embryo, will differentiate into either neurons or macroglia. Although the contributions of Germanic scientists predominated in the nineteenth century, at the beginning of the twentieth century, the so-called "Spanish histological school" played an important role.

Cajal's great virtue, which he succeeded in implanting in his successors, was that he was able to interpret what he saw and make schematic drawings combining what would be seen in the transverse and longitudinal planes in the same image. In fact, another Spaniard, Pío del Río Hortega, proposed that oligodendrocytes were the source of myelin in the central nervous system and compared oligodendroglial cells with Schwann cells, concluding that both were related to myelin sheaths. He also proposed the name microglia and their immune, migratory, and phagocytic functions based on previous studies published in 1900 by William Ford Robertson and Frantz Nissl, who described a type of activated glia in the form of rods, and in 1910 by Jean Nageotte, who described secretory granules in glial cells of the grey matter. Today, it is known that microglial cells or microglia also come from neuroepithelial cells of embryonic origin, specifically from fetal macrophages that penetrate the neural tube in early embryonic stages. In their resting, inactive state, they remain in a branched form. If microglia are activated, they acquire an amoeboid morphology, become more rounded, increase in size and acquire typical macrophage functions such as phagocytosis, and release of proinflammatory substances.[4] Microglia maintain their mitotic capacities in adulthood; that is, they continue to divide, although at a very slow rate. Together with astrocytes, they are mainly responsible for the pathoplasticity of the nervous system in conditions of chronic pain: they activate, multiply, and promote the release of proinflammatory substances that will stimulate the activation of other glial and neuronal cells, promoting and magnifying the transmission of information (Fig. 1).

However, science also caters to "fashions", and with the *neuronal doctrine*, everything began to revolve around the neuron and synaptic connections. The development of new physiological techniques once again increased the importance

---

[4] Microglial cells have different functions: they function as phagocytic cells that engulf the cell debris resulting from neuronal death mainly during development; they are also capable of inducing cell death of certain neurons; they have an immune role and have markers on their surface that recognize molecules that patrol the environment. It is estimated that apart from this resident microglia in the central nervous system, there is another type of invasive microglia called perivascular mononuclear phagocytes that cross the blood–brain barrier when there is an insult to the central nervous system.

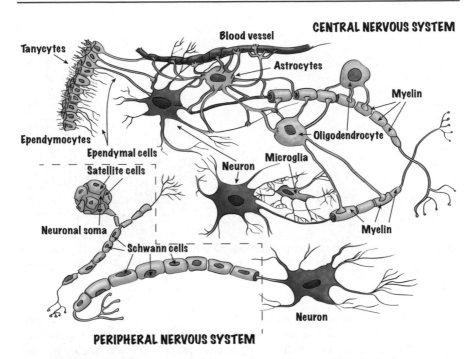

**Fig. 1 Different glial cells are present in the nervous system**. Anatomically, the nervous system is divided into the central and peripheral nervous systems. The types of glial cells that support neurons differ according to whether they are found in each of these parts. In the central nervous system, glia consist of: astrocytes, the most numerous glial cell type and important regulators of homeostasis and metabolism; oligodendrocytes, cells that form the myelin sheaths that wrap neuronal axons; ependymal cells, including tanycytes, which cover the cerebral ventricles and the central canal of the spinal cord, and are very important for the maintenance of cerebrospinal fluid; and microglia, cells with immune functions. On the other hand, the glia of the peripheral nervous system include: Schwann cells with a function analogous to that of oligodendrocytes; they cover neuronal axons, and can form myelin sheaths. Satellite cells—small cells that surround neuronal bodies forming a kind of capsule—are found at the interface between the peripheral and central systems in the dorsal root ganglia

of neurons, and glia, which were once again relegated to a supporting and nutritional role. It has not been until the last few decades that a revolution or rethinking of the neuronal doctrine has taken place, with glia going from being mere passive cells with trophic or neuronal support functions to major players. A search in the main scientific databases clearly illustrates that research on glia has increased exponentially since the 1960s (Fig. 2); and since then, there have been numerous publications claiming that glial cells make up 90% of the brain, while neurons make up the remaining 10%. Several authors have determined that 16% of nervous system cells in nematodes are glial cells, 20% in flies, 60% in rodents, 80% in chimpanzees, and 90% in humans. That is, the neuron:glia ratio in the human nervous system is 1:9 in favor of glia.

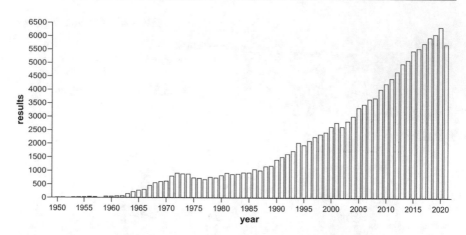

**Fig. 2 Evolution of the number of publications on glia in the last 70 years.** The number of results per year since 1950 is shown using the search terms *neuroglia OR glia* in PubMed (https://www.ncbi.nlm.nih.gov/pubmed/)

In a logical reasoning, the scientific community has deduced that something that presents such abundance must not have trivial functions, having spread the idea that the brains of more evolved organisms present a greater number of glial cells. That is, the ratio of glia to neurons increases as our definition of intelligence increases. In fact, it has been proposed that not only does the ratio of glia to neurons increase over the course of evolution, but so does the size of the glia themselves. This greater presence of glia in humans than in other animals could explain why humans are more susceptible to developing neurodegenerative diseases. Alzheimer's disease, amyotrophic lateral sclerosis (ALS), some migraines and epilepsies, psychiatric diseases, infectious diseases, Alexander's disease, cerebral infarctions, meningitis, and chronic pain are some of the diseases in which glial malfunctioning seems to play a fundamental role. In addition, approximately 80% of head tumors are gliomas.

However, as mentioned, not everything that is known about glia has been discovered in the last 60 years, nor has it all been ostracized during the second half of the 19th and first half of the twentieth century. What has been determined in recent years is that glial cells are probably the most versatile cells in the organism: they promote neuronal survival, release neurotrophic factors, phagocytize waste products in the extracellular milieu, release vasoconstrictors and vasodilators, are suspected of playing a regulatory and/or insulating role at neuronal synapses in communication that passes from one neuron to another, and so on. However, much remains to be discovered, and it is thought that there are numerous subtypes for the different glial cells discovered thus far.

## Not Everything is What It Seems

Historical perspective is fundamental to understand how certain questions come to be posed and how we try to respond to them. In 2005, a Brazilian researcher, Suzana Herculano-Houzel, noticed that most of the papers published on glia established a higher number of glia than neurons in the central nervous system: 100 billion neurons *vs.* 10 times more glial cells, based mainly on review papers, and not on original experimental articles. She decided to pull the thread, and followed the articles referenced in the review articles until she reached the original papers: a couple of experimental articles published in the 1960s in which the number of glial cells and neurons in a region of the cerebral cortex had been counted at random. With the help of a microscope, the number of neurons had been counted in a fixed image of a specific part of the brain, and that had been extrapolated to the whole brain. This is how they reached the conclusion that only 10% of the brain was made up of neurons. Could it really be that we only use a tenth of our brain, or is there a different neuron:glia ratio in other parts of the brain?

Dr. Herculano-Houzel developed a simple, relatively inexpensive and quick technique to quantify the number of total neurons in the brain: the isotropic fractionator (Fig. 3). Basically, the technique consisted of preserving the brain in a fixation solution that preserved and hardened the tissues. Once fixed, the brain (or alternatively, different parts of it) was homogenized in a detergent solution by mechanical action, as if mashed in a mortar. Since cell membranes are essentially made up of lipids, the use of a detergent would help to breakdown the tissues and cells, releasing the materials contained within. The resulting solution was centrifuged to obtain two perfectly delimited phases: a liquid phase, the supernatant, where the soluble remains of the broken cells would remain; and a precipitate in which practically only the cell nuclei would be found. Thanks precisely to the previous fixation step to which the brain was subjected, the nuclei, hardened, would have been able to withstand the mechanical action during the homogenization step. Returning to the phase separation, discarding the supernatant, and resuspending the sediment in a new volume would allow only nuclei to be retained in a homogeneous solution. The next step would be to take a small aliquot of the entire volume and add a blue fluorescent nuclei marker called DAPI. All nuclei in the solution would now be colored blue. By using a marker (NeuN)[5] that is able to recognize and bind specifically to the nuclei of neurons, only nuclei belonging to neuronal cells could be stained with red fluorescence.

Finally, with the help of a *Neubauer chamber*, or *hemocytometer* (a thin, flat piece of glass with drawn grids that can hold 0.1 µl every 16 squares), the number of nuclei that emit red fluorescence can be counted, revealing the number of neuronal nuclei present in 0.1 µl. Extrapolating to the initial volume in which the

---

[5] NeuN is a conjugated antibody, a protein with a coupled marker that binds tightly to a specific biological molecule in a sample, such as a nuclear protein. In this way, nuclei to which this antibody has bound can be detected by the action of the marker it carries, emitting a red fluorescent signal.

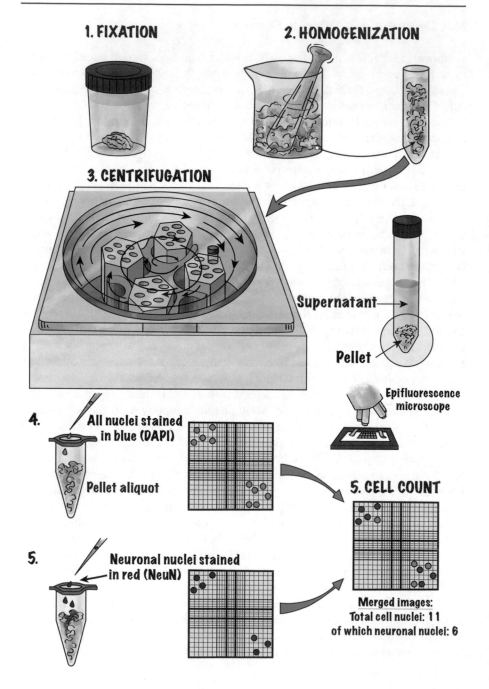

**1. FIXATION**

**2. HOMOGENIZATION**

**3. CENTRIFUGATION**

Supernatant

Pellet

Epifluorescence microscope

**4.** All nuclei stained in blue (DAPI)

Pellet aliquot

**5. CELL COUNT**

**5.** Neuronal nuclei stained in red (NeuN)

Merged images:
Total cell nuclei: 11
of which neuronal nuclei: 6

◄**Fig. 3   Isotropic fractionator method and nuclear labelling**. Dr. Herculano-Houzel devised the isotropic fractionator method to know the number of neurons in the brain as a whole and in each part of the brain, realizing the error of assuming that a small randomly chosen area is a reliable representation of what is happening in the rest of the brain. The technique consists of fixing the brain with a fixative solution that preserves the tissue before homogenizing it mechanically using a detergent. The homogenate is centrifuged to separate a liquid phase (the supernatant) from a solid phase (the precipitate), which contains mostly cell nuclei. Despite the mechanical friction during the homogenization step, the nuclei will remain intact thanks to the previous fixation step, which gives them a certain hardness. The precipitate containing the nuclei was dissolved in a volume to be by the investigator. The nuclear marker DAPI, blue in color, is added to an aliquot of this solution to observe all the cell nuclei present. In addition, the marker NeuN was added, which exclusively stains the neuronal nuclei red. Using a *Neubauer chamber*, the cell nuclei can be observed under a fluorescence microscope: in blue, all the nuclei present in the sample; and in red, only the neuronal nuclei. Therefore, the neuron/glia ratio can be easily calculated in the brain or, alternatively, in different parts obtained from the brain

nuclei had been diluted will reveal how many neurons are in the brain. To find the number of non-neuronal cells, simply subtract the number of neurons (red fluorescence) from the total number of cells emitting blue fluorescence, and extrapolate to the total volume in the same way. Considering that brains are mainly made up of neuronal cells and glia cells—it is estimated that the representation of endothelial cells and fibroblasts within the brain is less than 1%—a neuron:glia ratio in the brain can be calculated that is very close to the real figure.

Thanks to this technique, Dr. Herculano-Houzel confirmed that there are equal numbers of neuronal and non-neuronal cells in the human brain: 86,100 trillion neurons versus 84,600 trillion non-neuronal cells. That is, roughly one neuron for every glial cell. Contrary to what had been thought up to that time, glial cells are not the major cell type in the brain. She did not stop there, but began to perform this technique on the brains of different mammalian species. This neuron:glia ratio remained close to 1:1 in all the species she studied, reaching the conclusion that the number of neurons and glial cells followed the same pattern even among different animal species separated in evolution 90 million years ago.

The brain of a mammal can be peeled like a tangerine so that different parts (cerebral cortex, olfactory bulb, cerebellum, and the rest of the brain) can be separated relatively easily. Without the need to make incisions or cuts that damage the tissue, an uncontaminated separation between areas can be ensured. Analyzing these regions separately and comparing different species, Dr. Herculano-Houzel observed that the human cortex had many more neurons than the cerebellum, curiously just the opposite of what happened in the elephant. In fact, despite the difference in size between the two brains—the brain of an elephant is approximately three times larger than that of a human—the cerebral cortex of the African elephant had three times fewer neurons than that of a human.

Although overturning an idea that remains rooted in society or a routine methodology of research laboratories is enormously complex; the scientific method fortunately serves to subject hypotheses to the most effective refutation, even when they have been with us for generations. No, it is not true that we only use 10% of our brain, nor that, unlike other mammals, our brains have a ratio of glial cells to neurons of 10 to 1. Suzana Herculano-Houzel's discovery reflects that, for years, neuroscience (and various biomedical disciplines) had been relying on erroneous conclusions. Hence, the importance of the scientific method and preclinical research to resolve fundamental neuroscience questions is still unknown.

# Bibliography

Azevedo FA, Carvalho LR, Grinberg LT, Farfel JM, Ferretti RE, Leite RE, Jacob Filho W, Lent R, Herculano-Houzel S (2009) Equal numbers of neuronal and nonneuronal cells make the human brain an isometrically scaled-up primate brain. J Comp Neurol 513(5):532–541

Batelli S, Kremer M, Jung C, Gaul U (2017) Application of multicolor FlpOut technique to study high resolution single cell morphologies and cell interactions of glia in *Drosophila*. J Vis Exp 128:56177

Herculano-Houzel S, Lent R (2005) Isotropic fractionator: a simple, rapid method for the quantification of total cell and neuron numbers in the brain. J Neurosci 25(10):2518–2521

Herculano-Houzel S (2016) The human advantage: a new understanding of how our brain became remarkable. The MIT Press. ISBN: 9780262034258

Herculano-Houzel S (2009) The human brain in numbers: a linearly scaled-up primate brain. Front Hum Neurosci 3:31

Herculano-Houzel S (2014) The glia/neuron ratio: how it varies uniformly across brain structures and species and what that means for brain physiology and evolution. Glia 62(9):1377–1391

Herculano-Houzel S, Avelino-de-Souza K, Neves K, Porfírio J, Messeder D, Mattos Feijó L, Maldonado J, Manger PR (2014) The elephant brain in numbers. Front Neuroanat 8:46

Koob A (2019) The root of thought. Pearson Education Inc., New Jersey

Lent CM (1977) The Retzius cells within the central nervous system of leeches. Prog Neurobiol 8(2):81–117

Ndubaku U, de Bellard ME (2008) Glial cells: old cells with new twists. Acta Histochem 110(3):182–195

Stout RF Jr, Verkhratsky A, Parpura V (2014) Caenorhabditis elegans glia modulate neuronal activity and behaviour. Front Cell Neurosci 8(67):1–9

Verkhratsky A, Butt A (2007) Glial neurobiology. A textbook. Wiley, Chichester (UK)

Verkhratsky A (2006) Patching the glia reveals the functional organization of the brain. Pflugers Arch 453(3):411–420

**Miguel M. Garcia** holds a degree in Biochemistry from Universidad Autónoma de Madrid (UAM), a master's degree in the Study and Treatment of Pain, and a Ph.D. in Pain Research from Universidad Rey Juan Carlos (URJC). He is currently Assistant Professor and Researcher in the Area of Pharmacology, Nutrition and Bromatology of the Department of Basic Health Sciences at Universidad Rey Juan Carlos. He belongs to the High Performance Research Group in Experimental Pharmacology (PHARMAKOM) and coordinates the Teaching Innovation Group in Diseases and their Treatment (EducaPath) of Universidad Rey Juan Carlos. He is a member of the International Association for the Study of Pain (IASP), the Spanish Pain Society (SED) and the Spanish Society of Pharmacology (SEF). His research work focuses on basic pharmacology in the field of pain, particularly on the role of glial cells and cannabinoid and TLR4 receptors in nociception. As a disseminator, he has experience as a speaker at different science festivals and activities.

**Marina Martín Taboada** has a degree in Health Biology from Universidad de Alcalá de Henares (UAH) and a master's degree in Biochemistry, Molecular Biology and Biomedicine from Universidad Complutense de Madrid (UCM). She is currently a researcher in the Area of Biochemistry and Molecular Genetics of the Department of Basic Health Sciences at Universidad Rey Juan Carlos (URJC), where she is working on her Ph.D. thesis. She belongs to the High Performance Research Group in the Study of the Molecular Mechanisms of Glycolipotoxicity and Insulin Resistance: Implications in Obesity, Diabetes and Metabolic Syndrome (LipoBeta) of Universidad Rey Juan Carlos and her research work is focused on the study of obesity as a risk factor for conditions such as diabetes and chronic kidney disease in animal models and humans. In addition, she actively participates in research projects in collaboration with Hospital 12 de Octubre in Madrid. She is a member of the Spanish Obesity Society (SEEDO) and the Spanish Society of Biochemistry and Molecular Biology (SEBBM) and over the past years she has been participating in the organization of scientific dissemination activities such as *Science Week and the International Day of Women and Girls in Science.*

# Cajal's Cramps

Miguel Molina Álvarez◉ and Gema Vera Pasamontes◉

The idea is not new: what if small worlds existed within much larger worlds? In other words, what if the Earth is to our universe as a cell is to the human body; or a civilization created its world in a speck of dust and lived oblivious to the existence of the much larger world to which that speck belonged (like the inhabitants of *Whoville* in the children's book and animated film *Horton Hears a Who!*)? The concept can help us understand how the different cells of our body work and interact with each other. The cell sometimes acts as a huge logistics enterprise, with external communication networks (*extracellular signaling pathways*) and internal feedback networks (*intracellular signaling pathways*). These intracellular signaling pathways respond primarily to external stimuli, demanded by the client or society,

M. Molina Álvarez · G. Vera Pasamontes (✉)
Area of Pharmacology, Nutrition and Bromatology, Department of Basic Health Sciences, Unidad Asociada de I+D+i al Instituto de Química Médica (IQM-CSIC), Universidad Rey Juan Carlos, Madrid, Spain
e-mail: gema.vera@urjc.es

M. Molina Álvarez
e-mail: miguel.molina@urjc.es

M. Molina Álvarez
High Performance Experimental Pharmacology Research Group (PHARMAKOM), Universidad Rey Juan Carlos, Alcorcón, Spain

M. Molina Álvarez · G. Vera Pasamontes
Grupo Multidisciplinar de Investigación y Tratamiento del Dolor (i+DOL), Alcorcón, Spain

G. Vera Pasamontes
High Performance Pathophysiology and Pharmacology of the Digestive System Research Group (NeuGut), Universidad Rey Juan Carlos, Alcorcón, Spain

© The Author(s), under exclusive license to Springer Nature Switzerland AG 2024
M. M. Garcia (ed.), *Tales of Discovery*, https://doi.org/10.1007/978-3-031-47620-4_13

through signaling agents (messengers) that reach the cell surface to interact with a series of receptors integrated in that surface (the plasma membrane).

Receptors are responsible for transmitting external demands from the outside to the inside of the cell. But to get a response, as in large companies, many intermediaries, the second messengers, are needed. Sometimes, due to their characteristics, signaling agents can also diffuse through the cell surface and transmit the message directly to the cellular interior, without the need for intermediaries. They have "VIP authorization." This is the case with nitric oxide. Due to its small size and its gaseous state, nitric oxide easily crosses the plasma membrane to interact directly with its final receptor inside the cell: a soluble guanylyl cyclase in the cytoplasm. Nitric oxide is present in the whole organism but depending on the tissue in which it is found, its receptor will produce different responses. That is, the messenger is the same, but the message transmitted is different depending on the "company" that receives it. For example, in endothelial tissue, nitric oxide acts as a vasodilator, and in the central nervous system, it acts as a neurotransmitter. Fortunately, it is an unstable molecule, having a half-life of 6 s and diffusing a maximum of only 400 microns from the cell where it is released. Its action occurs close to its place of emission, thus preventing it from producing undesirable effects in other tissues or parts of the body.

Nitric oxide is able to interact with certain proteins responsible for receiving the message and which act accordingly by modifying its activity. However, the most studied role of nitric oxide is its ability to stimulate soluble guanylyl cyclase proteins in the cytoplasm, as discussed above. This pathway is better known as the *cGMP signaling pathway*. The different responses produced by binding the "messenger" nitric oxide to its receptor inside the cell, and the consequent activation of cGMP, involve various changes at the cellular and physiological levels: smooth muscle relaxation, synaptic plasticity, cardiac hypertrophy, modulation of the immune response, arteriolar vasodilatation, skeletal muscle perfusion, etc. Nitric oxide is also involved in the regulation of sexual health, sleep, memory, digestion, and pain. Among all these processes, we will address its role in gastrointestinal smooth muscle relaxation, which is directly related to digestion and indirectly related to visceral pain.

## Role of Nitric Oxide in the Enteric Nervous System

Signal transduction initiated by nitric oxide sets in motion a chain of actions that, in the digestive tract, ultimately leads to inhibition of enteric neurotransmission. Inhibition can act both on information of cessation and initiation of gastrointestinal motility. What does this mean? Well, by inhibiting the activity of the motor neurons of the digestive system, motility is reduced, and that makes nitric oxide an attractive pharmacological target for the control of intestinal discomfort and visceral pain. In the late 1980s, immunohistochemical studies and pharmacological analyses (tests in which drugs of known action are used to study their effect on tissues, organs, or whole organisms) showed that enteric neurons (i.e., those found in the enteric nervous system between the muscular layers of the digestive tract), either excitatory or inhibitory of muscular contractile activity, contained various types of transmitters. The (neuro) transmitters (discussed in another chapter), would be like the baton passed by the runners in a relay race or, in our analogy, like the message passed by the different messengers in a company's courier service until it reaches its addressee. This gave rise to the *chemical code hypothesis*. This suggests that due to a combination of specific chemical markers, "companies" (i.e., the different neuronal cells) could be recognized and differentiated from their peers not by their appearance but by their tasks or functions. However, the problem is somewhat more complex since certain transmitters are present in all types of neurons located in the different regions of the gastrointestinal tract; that is to say that there are functions shared by all companies.

Let us take a closer look at what happens in the enteric nervous system. The digestive tract is made up of a series of epithelial cells that cover the digestive tract both externally and internally to protect it from possible threats and aggressions, such as microorganisms or the transit of food. These layers would serve as the enclosure that protects the companies from intruders. The epithelial cells form the so-called *serous layer* (outer face) and *mucous layer* (inner face). Since it needs to protect itself from those microorganisms that manage to circumvent this first barrier, the digestive tract also has an extensive immune system, which could be analogous to "security agents" inside the enclosure. Because it needs to move food (alternatively, waste) along the digestive tract, it also has abundant musculature and a nervous system of its own: the enteric nervous system.[1] The transmission of commands to the musculature can occur directly, by innervation of motor neurons; or indirectly, by controlling the excitability of other nonnerve cells called *interstitial cells of Cajal*,[2] intimately interconnected with muscle cells.

---

[1] The longitudinal musculature envelops and exerts pressure on a second circular muscle. The first is responsible for the forward movement of the food bolus; the second for the peristaltic movements that turn the food around so that all of it comes into contact with the walls of the tube, and all of the food can be absorbed. For its part, the enteric nervous system is a subdivision of the autonomic nervous system that is in charge of controlling the digestive system; as its name indicates, its activity is autonomous, involuntary. The enteric nervous system is divided into two plexuses: the myenteric plexus and the submucosal plexus. The former innervates the musculature described above, being in charge of gastrointestinal movements. The latter regulates the secretion of various substances along the digestive tract.

[2] In 1889 Santiago Ramón y Cajal described for the first time new cells that did not conform to any of the previously known cells, which he considered to be primitive accessory neurons. From his discovery until his death, Ramón y Cajal defended that these cells were a different class of neurons, although curiously they did not share the morphology of the rest of the neurons he had previously described (dendrites, soma and axon). He called this new type of cells *interstitial cells*. In the following years there was a great controversy among prominent scientists of the day, because with the methods and instruments they had at that time it was not possible to determine what type of cells they truly were. In 1895, the Russian neuroscientist Alexander Stanislavovich Dogiel first named interstitial cells as *Cajal's interstitial cells*, in honor of the Spanish scientist's discovery. It was Nobel Prize winner Sir Arthur Keith who first proposed, in 1915, that perhaps these cells could function as a pacemaker system in the digestive tract and was the first to propose the possible

These interstitial cells of Cajal act as if they were a pacemaker, regulating the contractility of muscle fibers in a rhythmic and organized way, similar to the tides of the sea. They generate slow waves of ions ($Ca^{2+}$) that can spread both from different compartments of the cellular interior, which function as reservoirs, and from the cellular exterior, occupying the entire cytosol, and culminating in muscular contraction. The accumulation of positive charges on the internal face in direct contact with the cell surface generates a change in polarity in the cell, making it electrically excitable. A voltage change occurs on both sides of the plasma membrane.

The control of Cajal activity can be carried out both by neuronal stimuli and by hormonal stimuli that trigger this *cytosolic calcium oscillator mechanism.* In this respect, nitric oxide is able to hyperpolarize the membrane of these cells, making them less susceptible to excitation. This is achieved by decreasing the activity of channels responsible for introducing calcium into the cell from outside the cell (L-type channels),[3] and as mentioned above, the entry of calcium into the cell interior is essential for the propagation of the nerve impulse. The end result is inhibit transmission of the nerve impulse to the neighboring smooth muscle cells and, consequently, block their contractile activity. In other words, nitric oxide negatively modulates the pacemaker activity of the interstitial cells of Cajal (Fig. 1).

The interstitial cells of Cajal (ICC) are distributed as a network occupying characteristic positions along the entire digestive tract. They adopt different patterns of distribution and morphology according to the anatomical location they occupy and can be classified into different subtypes: ICC of the subserosa; ICC of the longitudinal muscle; ICC of the myenteric plexus; ICC of the circular muscle; and ICC of the submucosa in the stomach, of the deep muscle plexus in the small intestine, or of the submucosal plexus in the colon (Fig. 2). Therefore, these Cajal cells are not only in contact with cells from the smooth musculature but are also with epithelial and neuronal cells.

That said, both excitatory and inhibitory motor neurons release neurotransmitters that act on Cajal cells, which are coupled to smooth muscle cells through special junctions that connect the cytoplasm of one cell to the other, called *slit* or *communicating junctions.* This is why transmitters released from neurons can

---

muscular nature of these cells. However, it was not until the 1960s when histological techniques were developed that were finally able to discern the nature of Cajal's interstitial cells, proving Sir Arthur Keith right 50 years later.

[3] Evidence that nitric oxide is an enteric inhibitory transmitter came with the work of Rand and Sanders & Ward in 1992. It is now known to act either autocrine as a second messenger in the cell of origin itself, or it can diffuse across membranes and act in neighbouring cells as a paracrine signalling agent. Nitric oxide is synthesized from arginine and oxygen by any of the three isoforms of the enzyme nitric oxide synthase (NOS), which are named after the tissues where they were first described or the pathway by which they are controlled (nNOS, neuronal; iNOS, inducible or immunocytic; and eNOS, endothelial). However, the expression of these enzymes is not restricted to what their names indicate, but they are widely expressed and can coexist in many cell types. However, the regulation of NOS is very different for the three isoforms and depends partially on where they are located in the cell.

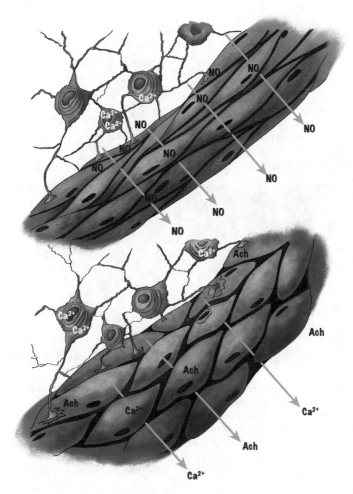

**Fig. 1** Arrangement of neurons and muscle fibers in the myenteric plexus or Auerbach's plexus. It shows the interaction between neurons and muscle fibers in the myenteric plexus and the Ca flux$^{2+}$ responsible for pacemaker activity in the interstitial cells of Cajal. Nitric oxide (NO) is a short-acting and rapidly diffusing messenger that can either diffuse from other cells or be generated within the neuron itself to stimulate the production of a second messenger (cGMP), specifically, which will promote the entry of $Ca^{2+}$ into the cytoplasm. In this way, from NO, the interstitial cells of Cajal are able to generate periodic pulses of $Ca^{2+}$ which start as bursts that activate other channels, including the $Cl^-$ channels. The opening of these channels generates a spontaneous transient current of $Cl^-$ ions$^-$ towards the interior (greater negative charge inside the cell, which is *hyperpolarization*). Therefore, NO will end up acting as a brake or autoregulator of the action of the pacemaker cells. On the other hand, the gradual increase in the concentration of $Ca^{2+}$ in the cytosol results in the accumulation of transient currents and produces the spontaneous depolarization of the cell; it also propagates towards the neighboring smooth muscle cells through slit junctions. This depolarization of the membrane of these muscle cells will produce their contraction

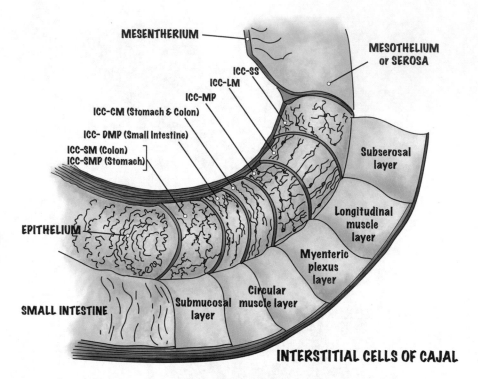

**Fig. 2**  Different types of interstitial cells of Cajal according to their location in the muscular layers and the different organs of the gastrointestinal tract. Interstitial cells of Cajal of the circular muscle (*ICC-CM*); interstitial cells of Cajal associated with the deep muscular plexus (*ICC-DMP*); interstitial cells of Cajal of the longitudinal muscle (*ICC-LM*); interstitial cells of Cajal associated with the myenteric plexus (*ICC-MP*); interstitial cells of Cajal associated with the submuscular plexus in the colon (*ICC-SM*); interstitial cells of Cajal associated with the submuscular plexus in the stomach (ICC-SMP); interstitial cells of Cajal associated with the subserosa membrane (*ICC-SS*)

also diffuse and act directly on muscle cells. Since the transmission carried out by inhibitory motor neurons could not be described by any of the peripheral transmitters known thus far, this type of transmission was called *nonadrenergic, noncholinergic* (NANC) transmission concerning motor neurons inhibiting gastrointestinal smooth muscle contraction. All these inhibitory motor neurons specifically release nitric oxide, adenosine triphosphate (ATP), peptides of the vasoactive intestinal peptide (VIP) family, and pituitary adenylyl cyclase activating peptide (PACAP). In other words, they can be characterized by their *chemical code*. All these neurotransmitters released from the endings of these neurons have inhibitory actions on the cells of Cajal and neighboring muscle cells, so all these "messengers" would produce the same response in all "companies": inhibition of muscle contraction.

However, not all of these neurotransmitters contribute to muscle relaxation in the same proportion but vary according to the region of the gastrointestinal tract on which they act, and the animal species under study. The fact that these neurons present multiple transmitters led us to hypothesize (correctly) that a compensation system could exist to prevent the risk of deficient control of the inhibition of digestive motility, which could put the organism in clear danger. Back to our analogy, imagine that an important messenger is ill and cannot go to work. Fortunately, there are others who that can replace that messenger's function, ensuring that the message reaches its addressee. Likewise, although nitric oxide seems thus far to be the most important agent, several studies have demonstrated the viability of mutant mice lacking the enzyme that generates nitric oxide in the nervous system (nNOS), being perfectly fertile and lacking histopathological abnormalities in the central nervous system; this would confirm the hypothesis.

## The Enteric Nervous System Beyond Gastrointestinal Movement

We have already seen that the tissues of the body use nitric oxide as a messenger for different commands; now let us look specifically at what happens in neurons with this messenger. Neurons are the cells with the highest levels of nitric oxide in the whole organism. Although we have previously mentioned that mutant mice lacking nNOS are perfectly viable, this does not mean that the absence of correct regulation of NOS activity in the nervous system can contribute to or give rise to numerous pathologies in other species.

Before we go on, some backstory: In 1990, several researchers revealed that NOS inhibitors and nitric oxide scavengers reduced inhibitory transmission on gastrointestinal smooth muscle; that is, gastrointestinal activity was significantly increased. For this reason, in the last decade, efforts have been made to understand the role of nitric oxide in spinal afferents[4] for the study and treatment of pain.

In the plasma membrane of neurons, as in all cells of the body, a number of receptors are present. Within this group of receptors, there is a class that is coupled to ion channels, but how do they work? When the neurotransmitter (the messenger) binds to its corresponding receptor, it transmits a command. In this case, there are only two possible commands: open the channel or close it. The opening or closing of the channel will produce changes in the cell relative to its excitability and, consequently, will modify nerve transmission. Recently, the importance of the role of the TRPV2 receptor has been demonstrated.[5] This receptor is able to respond proportionally to noxious stimuli, such as spontaneous contractions or cramps, and

---

[4] Spinal afferents are all the sensory signals, nociceptive and nonnociceptive (tactile, thermal, or proprioceptive information), that reach the spinal cord from the nerves of the periphery. This is the meeting point of the peripheral nervous system and the central nervous system.

[5] Transient receptor potential (TRP) channels belong to a group of receptors coupled to ion channels that detect thermal, chemical, and mechanical stimuli. Specifically, TRPV2 are expressed in

increase the synthesis of nitric oxide through the activation of nNOS, thus reducing the excess activity; consequently, it is able to reduce the symptomatology.

But what happens with nitric oxide in the central nervous system? Thus far, we have seen what happens at the peripheral level, but what happens at the central level is no less interesting. In 2007, a series of experiments with dogs revealed that nitric oxide plays a fundamental role in the modulation of pain. Most of us have seen a war movie or an action film in which someone keeps running despite being hit by a bullet in the leg or arm. Is that true or just a dramatic touch in a fictional film? Indeed, this can happen in real life. The brain has involuntary pain modulating mechanisms, and our messenger, nitric oxide, contributes to them, and is fundamental to activating the pain inhibitory mechanism. In fact, it has been proven that when the activation of a descending inhibitory mechanism of pain is reduced due to the absence or blockage of nitric oxide in the central nervous system, pain (in this case visceral) is favored, and may become chronic.

Therefore, it seems that the solution is obvious. If nitric oxide helps to reduce pain both peripherally and centrally, a drug could be synthesized to ensure sufficient nitric oxide throughout the nervous system. However, as is often the case, nothing is that simple. Studies conducted for that purpose concluded that an excess of nitric oxide in the body can be toxic and lethal to cell viability. As we mentioned above, although it has a short diffusion power, nitric oxide is able to cross the plasma membrane of any cell, which could produce numerous responses within neurons. That makes it very difficult, if not impossible, to control—especially if administered at will. Moreover, we must emphasize the complexity of visceral pain: it is a nonspecific, dull pain; it is difficult to describe; and it is sometimes perceived as being felt in parts of the body other than its actual source, e.g., in the case of myocardial infarction, it is common for the patient to report pain in the left arm. It should be noted that nitric oxide-releasing drugs have found their place in clinical practice (although not for the purpose of alleviating visceral pain), due to their hypotensive effect, which allows vasodilation and prevention of heart disease.

The sociohistorical context of the Spain in which Santiago Ramón y Cajal lived was marked by a period of decadence at all levels, most profoundly illustrated by the *Disaster of '98*.[6] His work takes on an even greater transcendental value because of the complexity he had to face as a scientist developing his career in that Spain. As he would say, "the cart of Spanish culture lacks the wheel of science."

---

gastric and intestinal enteric neurons. Their activation relaxes the stomach and intestinal musculature, and promotes gastrointestinal transit.

[6] In 1898, the United States intervened in the Cuban War of Independence (1895–1898) beginning the Spanish-American War (from April to August 1898). During this war, Spain lost its last overseas colonies gaven rise in Spain to the expression "Desastre del 98" (Disaster of '98). Following its defeat, Spain lost Cuba, Puerto Rico, the Philippines and Guam, and the Pacific possessions were sold or mis-sold to Germany because Spain was unable to defend them.

# Biblography

Benavides-Trujillo MA, Pinzón-Tovar A (2008) Nitric oxide: pathophysiological implications. Rev Col Anest. 36:45–52

Berridge MJ (2008) Cell signalling biology. Portland Press, Lmtd, Portland

Bredt DS, Hwang PM, Snyder SH (1990) Localization of nitric oxide synthase indicating a neural role for nitric oxide. Nature 347:768–770

Bredt DS (1999) Endogenous nitric oxide synthesis: biological functions and pathophysiology. Free Radic Res 31(6):577–596

Bult H, Boeckxstaens GE, Pelckmans PA, Jordaens FH, Van Maercke YM, Herman AG (1990) Nitric oxide as an inhibitory nonadrenergic noncholinergic neurotransmitter. Nature 345(6273):346–347

Burnstock G, Campbell G, Bennett MR, Holman ME (1964) Innervation of the guinea-pig taenia coli: are there intrinsic inhibitory nerves which are distinct from sympathetic nerves? Int J Neuropharmacol 3:163–166

Cantarero Carmona I, Junquera Escribano MC (director) (2011) Original contributions to the knowledge of Cajal's interstitial cells [doctoral thesis]. University of Zaragoza

Centelles JJ, Esteban Redondo C, Imperial S (2004) Nitric oxide: a toxic gas that acts as a blood pressure regulator. Offarm 23(11):96–102

Furness JB, Morris JL, Gibbins IL, Costa M (1989) Chemical coding of neurons and plurichemical transmission. Ann Rev Pharmacol Toxicol 29:289–306

Furness JB, Young HM, Pompolo S, Bornstein JC, Kunze WAA, McConalogue K (1995) Plurichemical transmission and chemical coding of neurons in the digestive tract. Gastroenterology 108(2):554–563

Furness JB (2006) The enteric nervous system. Blackwell Publishing, Massachusetts

Gibbins IL, Morris JL, Furness JB, Costa M (1987) Chemical coding of autonomic neurons. Exp Brain Res 16:23–27

Grozdanovic Z (2001) NO message from muscle. Microsc Res Tech 55(3):148–153

Hanani M, Farrugia G, Komuro T (2005) Intercellular coupling of interstitial cells of Cajal in the digestive tract. In Rev Cytol. 242:249–282

Hirst DG, Robson T (2011) Nitric oxide physiology and pathology. Methods Mol Biol 704:1–13

Koike S, Uno T, Bamba H, Shibata T, Okano H, Hisa Y (2004) Distribution of vanilloid receptors in the rat laryngeal innervation. Acta Otolaryngol 124(4):515–519

Komuro T (2006) Structure and organization of interstitial cells of Cajal in the gastrointestinal tract. J Physiol 576(Pt 3):653–658

Li CG, Rand MJ (1990) Nitric oxide and vasoactive intestinal polypeptide mediate nonadrenergic, noncholinergic inhibitory transmission to smooth muscle of the rat gastric fundus. Eur J Pharmacol 191(3):303–309

Mashimo H, Kjellin A, Goyal RK (2000) Gastric stasis in neuronal nitric oxide synthase-deficient knockout mice. Gastroenterology 119(3):766–773

Mihara H (2010) Involvement of TRPV2 activation in intestinal movement through nitric oxide production in mice. J Neurosci 30(49):16536–16544

Poole DP, Lieu T, Veldhuis NA, Rajasekhar P, Bunnett NW (2015) Targeting of transient receptor potential channels in digestive disease. In: Szallasi A (ed) TRP channels as therapeutic targets: from basic science to clinical use. Elsevier, London, pp 385–403

Rand MJ (1992) Nitrergic transmission: nitric oxide as a mediator of nonadrenergic, noncholinergic neuro-effector transmission. Clin Exp Pharmacol Physiol 19(3):147–169

Sanders KM, Ward SM (1992) Nitric oxide as a mediator of nonadrenergic noncholinergic neurotransmission. Am J Physiol 262(3 Pt 1):G379–G392

Sarna SK (2007) Enteric descending and afferent neural signaling stimulated by giant migrating contractions: essential contributing factors to visceral pain. Am J Physiol Gastrointest Liver Physiol 292(2):572–581

Soto-Abánades CI, Ríos-Blanco JJ, Barbado-Hernández FJ (2008) Cajal's interstitial cells: another contribution to modern medicine. Rev Clin Esp 208(11):572–574

Suthamnatpong N, Hata F, Kanada A, Takeuchi T, Yagasaki O (1993) Mediators of nonadrenergic, noncholinergic inhibition in the proximal, middle and distal regions of rat colon. Br J Pharmacol 108(2):348–355

Tan LL, Bornstein JC, Anderson CR (2008) Distinct chemical classes of medium-sized transient receptor potential channel vanilloid 1-immunoreactive dorsal root ganglion neurons innervate the adult mouse jejunum and colon. Neuroscience 156(2):334–343

Tan LL, Bornstein JC, Anderson CR (2009) Neurochemical and morphological phenotypes of vagal afferent neurons innervating the adult mouse jejunum. Neurogastroenterol Motil 21(9):994–1001

Tjong YW, Ip SP, Lao L, Wu J, Fong HH, Sung JJ, Berman B, Che CT (2011) Role of neuronal nitric oxide synthase in colonic distension-induced hyperalgesia in distal colon of neonatal maternal separated male rats. Neurogastroenterol Motil 23(7):666-e278

Toda N, Baba H, Okamura T (1990) Role of nitric oxide in nonadrenergic, noncholinergic nerve-mediated relaxation in dog duodenal longitudinal muscle strips. Japan J Pharmacol 53(2):281–284

Uludag O, Tunctan B, Altug S, Zengil H, Abacioglu N (2007) Twenty-four-hour variation of L-arginine/nitric oxide/cyclic guanosine monophosphate pathway demonstrated by the mouse visceral pain model. Chronobiol Int 24(3):413–424

Ward SM, Dalziel HH, Bradley ME, Buxton IL, Keef K, Westfall DP, Sanders KM (1992) Involvement of cyclic GMP in nonadrenergic, noncholinergic inhibitory neurotransmission in dog proximal colon. Br J Pharmacol 107(4):1075–1082

Yang Q, Underwood MJ, He GW (2012) Calcium-activated potassium channels in vasculature in response to ischaemia-reperfusion. J Cardiovasc Pharmacol 59(2):109–115

**Miguel Molina Álvarez** has a degree in Physical Therapy from Universidad Rey Juan Carlos (URJC), and a double master's degree in Manual Physiotherapy of the locomotor system from Universidad de Alcalá de Henares (UAH) and in Clinical and Basic Aspects of Pain from Universidad de Cantabria (UC)-Universidad Rey Juan Carlos. He currently combines his research work in animal models for the study and treatment of pain in the Area of Pharmacology, Nutrition and Bromatology of Universidad Rey Juan Carlos with his activity as a physiotherapist. He belongs to the High Performance Research Group in Experimental Pharmacology (PHARMAKOM) of Universidad Rey Juan Carlos and is a member of the Spanish Pain Society (SED).

**Gema Vera Pasamontes** holds a degree in Biology from Universidad Complutense de Madrid (UCM) and a Ph.D. in Pain Research from Universidad Rey Juan Carlos (URJC). She has conducted different internships in research centers at national and international levels and is currently an Associate Professor in the Department of Pharmacology, Nutrition and Bromatology of the Department of Basic Health Sciences of Universidad Rey Juan Carlos. She belongs to the High Performance Research Group in Gastrointestinal Pathophysiology and Pharmacology (NeuGut) of Universidad Rey Juan Carlos. Her research work focuses on the field of gastrointestinal motility and pain, particularly in gastrointestinal alterations derived from pharmacological treatments, using animal models. As a disseminator, she has participated in activities such as *Science Week*, the *European Researchers' Night* and the *European Brain Week*.

# CRISPR/Cas: Photoshopping DNA

Patricia Corrales Cordón◉ and Fernando Muñoz Muñoz

Like Henry Jones Jr. (better known as Indiana Jones—"Indy" to his friends), whose adventures to find the legendary Holy Grail were chronicled in the Hollywood film *Indiana Jones and the Temple of Doom*, scientists seek their own "Holy Grail" of genetic engineering. One of the most deeply rooted goals of every scientist is to apply the results of their laboratory experiments in everyday life, and improve the health of humans, animals,[1] or plants. And so, during recent decades, scientists have been compelled to study, learn, and understand appropriate tools designed to improve the mechanisms that lead to genetic modifications. They seek a tool capable of controlling the genetic alterations themselves, which in some cases lead

---

[1] DNA: Deoxyribonucleic acid, consisting of nucleotides known as Adenine, Guanine, Cytosine, and Thymine, as well as phosphate groups.

---

P. Corrales Cordón (✉)
Area of Biochemistry and Molecular Biology, Department of Basic Health Sciences, Universidad Rey Juan Carlos, Alcorcón, Spain
e-mail: patricia.corrales@urjc.es

High Performance Research Group in the Study of Molecular Mechanisms of Glucolipotoxicity and Insulin Resistance: Implications in Obesity, Diabetes and Metabolic Syndrome (LIPOBETA), Universidad Rey Juan Carlos, Alcorcón, Spain

Consolidated Research Group on Obesity and Type 2 Diabetes: Adipose Tissue Biology (BIOFAT), Universidad Rey Juan Carlos, Alcorcón, Spain

F. Muñoz Muñoz
Clinical Diagnostic Processes and Orthoprosthetic Products, Consejería de Educación de la Comunidad de Madrid, Madrid, Spain

© The Author(s), under exclusive license to Springer Nature Switzerland AG 2024
M. M. Garcia (ed.), *Tales of Discovery*, https://doi.org/10.1007/978-3-031-47620-4_14

to a disease so harmful to the living being that it can cause its death—or worse, its extinction.

After several years of research and hard work, this scientific quest has led to the improvement of a tool that has revolutionized genetic engineering (drum roll, please): the CRISPR system. CRISPR is the acronym for *Clustered Regularly Interspaced Short Palindromic Repeats*. While it is still under study, this sensational genetic tool allows scientists to precisely insert and modify specific DNA sequences in the genome of cells, without causing permanent alteration in all cellular genes. Thanks to the modifications in the DNA that generate this supergenetic tool (which seems a magic trick developed by scientists), the scientific world has hope that it can be used to cure diseases that include genetic alterations, or to improve food, or to develop biofuels that pollute significantly less than the current ones.

It is almost impossible to imagine that the researchers who discovered the CRISPR system—the Holy Grail of genetics—in 1987 were unaware of its potential and the repercussions this system would have, years later, on science in general, and on molecular biology in particular. This genetic tool is used by some organisms that lack even a microscopic brain, and are found in all kinds of environments, including bacteria. These organisms, which some people find unattractive, have been using the CRISPR system for thousands of years to protect their genome against attacks from pathogens such as viruses. Their protections generated an immune system capable of creating a kind of genetic memory that helps deal with the next attack. If we look at a phylogenetic tree, bacteria are among the oldest organisms that exist. This leads us to think that the genesis of their immune system took place billions of years ago. This is important because bacteria have been very strategic, despite not having a brain; they have retained these CRISPR sequences to survive and colonize other places around the planet. In fact, if they had not been able to conserve this genetic tool, they would have accumulated so many genetic mutations in their genome, due to the constant attacks against them, that they would have become extinct. How much we have to learn from these tiny organisms!

Let us return to the history of CRISPR. In 1987, the first scientific evidence was published by American researchers on the existence of a set of nucleotide sequence[2] (approximately 30 nucleotide base pairs), repeated and separated by spacer regions (approximately 36 nucleotide base pairs). These repeated sequences, together with the spacer regions, were described in the DNA of the well-known bacterium *Escherichia coli*. In 1993, on the other side of the pond, the Spanish researcher Francisco Mojica at the University of Alicante, discovered the DNA of the archaeon[3] *Haloferax mediterranei*. The DNA of this microorganism was special: it contained palindromic sequences. This means it reads the same way

---

[2] Nucleotides: chemical compounds made up of nitrogenous bases, sugars and phosphoric acid.

[3] Organisms formed by a single cell, without nuclei, similar to bacteria.

front-to-back or back-to-front, like the words "kayak" or "radar." These palin-dromic DNA sequences contained between 30 and 34 base pairs, and Mojica observed that they were arranged in tandem, and were separated by nonrepeat-ing sequences of 35 to 39 base pairs. Together, the palindromic sequences and the separator sequences were located in different regions throughout the chro-mosomal DNA of the archaeal *Haloferax*. At that time, it was thought that the repeated sequences were "junk" DNA—useless—and characteristic of this type of organism only. Years later, the palindromic sequences repeated in the genome of prokaryotic organisms[4] gained greater importance in the field of gene editing, as it was observed that this type of special sequence was widespread throughout the evolutionary tree. This led to the suggestion that they could be important for some evolutionary reason. Far from ignoring it, some researchers devoted their working hours (and very long free hours outside the workplace) to the study of the genome of microorganisms. They came to believe that these discoveries could result in one of the most important molecular and genetic techniques of the twen-tieth and twenty-first centuries. In 2020, Emmanuelle Charpentier and Jennifer A. Gouda were awarded the Nobel Prize in Chemistry for the development of this gene technology: the CRISPR/Cas9 genetic scissor or the photosshoping DNA.

Continuining with the depiction of CRISPR/Cas9 discovery, at the end of the 1990s DNA sequencing tools were developed to identify the nucleotide arrange-ment that gives rise to genetic information. Thanks to their use, new palindromic sequences, long and short, were discovered in tandem, interspersed in different microorganisms. As a direct result of all this research, in 2002, there was a con-sensus among various researchers around the world, dedicated to the study of these sequences, to adopt the acronym CRISPR (pronounced *crisper*) for these interspaced repeated sequences. However, it was not until 2005 when the works based on CRISPR sequences, were published and shared with the rest of the sci-entific community. All these publications led to the confirmation that the spacer sequences were complementary to the sequences of some types of viruses (such as "zombie" invasion) and attack bacteria, and that they were part of a molecule that shares prominence with CRISPR: the nuclease[5] Cas. Cas constitutes one addi-tional piece of the puzzle that was missing until this time. In fact, at the time the spacer sequences were discovered, Charpentier was working on the identification of additional regulatory RNAs in microbes—in fact, she was focused on the bio-chemical characterization of CRISPR. As for Goudna, she was immersed in the structural characterization of RNA–protein interactions. They met in a research conference and the two scientists decided to join forces to characterize the newly found machinery and repurposing it into a genome editing tool. Furthermore, we understand now that the spacer sequences of bacteria are equivalent to their immune memory and contain the history of all the attacks they have suffered from

---

[4] Organisms lacking a nucleus, and surrounded by a membrane.
[5] Enzymes that degrade nucleic acids.

different viruses and pathogens. What a genetic sophistication these tiny organisms have!

And yet, despite the eruption of the CRISPR volcano in the scientific world (numerous publications on the presence of this system in a multitude of prokaryotic organisms had been accepted in various scientific journals), the function of most of its components was still not known in detail. Thanks to the progress of bioinformatics analysis, scientists were gradually going deeper into the components that make up the CRISPR/Cas system, and were better able to understand the mechanisms that make this genetic adaptive defense system so important for prokaryotic organisms and for the field of genetic engineering. Finally, in 2013, the CRISPR/Cas system began to be used in genetic engineering in mammalian cells; it was already known that CRISPR was able to act by the activation of Cas. Since then, the study and analysis of the CRISPR/Cas system has continued to steadily grow; it is one of the most revolutionary and optimized techniques for genetic engineering. This tool has been used for the modification of genomes of multiple species, such as bacteria, plants, worms, zebrafish, rodents, and even human cells in culture. Over time, the CRISPR/Cas system became a very precise molecular tool capable of cutting and pasting a piece of DNA from the genome of cells. It has proven to have great scientific importance and relevance, and its biotechnological applicability is evident. In science, it is indispensable.

## The Magic of How CRISPR/Cas Works

Thus far we have only talked about how the CRISPR/Cas system was discovered, but the importance of this genetic technology lies in how it is able to generate this type of genetic immune system—as if it were conjured from thin air. Therefore, we will now focus on how the CRISPR works.

The CRISPR/Cas system, as mentioned above, is a genetic engineering tool based on the defense system of prokaryotic organisms. Some well-known bacteria, such as *Escherichia coli* or the archeon *Haloferax*, have developed a genetic adaptive immunity mechanism that gives them protection against the attack of viruses, plasmids, or other pathogens. In a first attack, the bacteria insert small DNA sequences of the attacker into the genome of the bacterium, thus generating a complete record of infections, so that before a second attack, the bacterium is protected. This is similar to what happens with the flu vaccines we get every year, except that bacteria have been protecting themselves from attack for millions of years in a genetic way that can be passed on from one generation to the next. We, on the other hand, are injected with a vaccine, and we generate antibodies against a specific pathogen to protect us against an attack by that pathogen; but, we are not able to genetically pass this protection to subsequent generations. In that respect, how the CRISPR/Cas system functions can seem like an extraordinary magic trick.

Can the functioning of the genetic discovery of the century be as simple as that: a trick? To understand the usefulness of this method, and the revolution it has generated in genetic engineering, we have to delve deeper and focus on the knowledge we have gained about how this genetic tool functions. Briefly, the external agent (e.g., a virus) attacks and injects its genetic material into the cytoplasm of the bacterium. The bacterium recognizes this external genetic material by the PAM (Protospacer Adjacent *Motif*), a short DNA sequence present on the genetic material of the invading virus but not found in the bacterial host genome. The external material incorporated into the bacterium genome corresponds to regions adjacent to PAM. However, as the attack is not carried out by a single pathogen, but by several, repeated sequences of this PAM-adjacent nucleotides are generated and placed in tandem. The newly incorporated material becomes known as CRISPR sequence (clustered regularly interspaced short palindromic repeats) and will serve as a template to detect future infections of a like pathogen. In fact, the entire bacterial genome would count on CRISPR sequences and other sequences repeated in tandem that, when transcribed into RNA receive the name of crRNA (*CRISPR-associated RNA*).

As just mentioned, this crRNA is always flanked by repeated sequences and, at this point, endonucleases, which are responsible for processing the crRNA, come into play. These molecules are known as Cas (*CRISPR-associated protein*, i.e., nucleases associated with the CRISPR sequences), are capable of being activated by the repeated sequences, and are the ones that cut the invading genome. When a second attack of the virus or external pathogen occurs, the process is much faster, and will generate the external genome cut specifically.

To simplify: a virus or pathogen attacks a bacterium. The virus injects its genome into the bacterium, which manages to introduce sequences from the attacker into its own genome. These sequences are interspersed between repeated sequences of the bacterium's genome in such a way that they form a tandem that is repeated several times throughout the bacterial genome. In the case of the CRISPR II system, the repeated sequence is transcribed into RNA, with which it forms a *dimer*. Thus, the CRISPR (repeat sequence)-spacer sequence forms a genetic *dimer* with tracrRNA (Fig. 1). This *dimer* binds to Cas9, and the CRISPR/Cas system is activated. In a second infection, the crRNA recognizes the PAM sequence, and Cas will cut them so that the genetic material of the second attacker does not come into contact with the bacterial genome, thus generating immunity to that specific virus or pathogen.

The CRISPR system is more than a genome editor; it is a tool capable of cutting and pasting genome sequences accurately, as does the text editor on our computer. Since the excision carried out by the CRISPR/Cas system will occur in both DNA strands of the bacterium affecting genome transcription, this cut must be repaired, either by directly rejoining the two fragments resulting from the cut or by inserting a repair template between them. However, we must be careful. Changes in the genomic sequence caused by the CRISPR/Cas system may have negative consequences and impair proper gene function.

Due to the precision with which the CRISPR/Cas system is able to cut at specific sites, this remarkably important genetics tool has turned out to be a research alternative to the classical methods of targeted genome editing. CRISPR/Cas has made it possible to insert, eliminate, or generate mutations in specific sequences. It

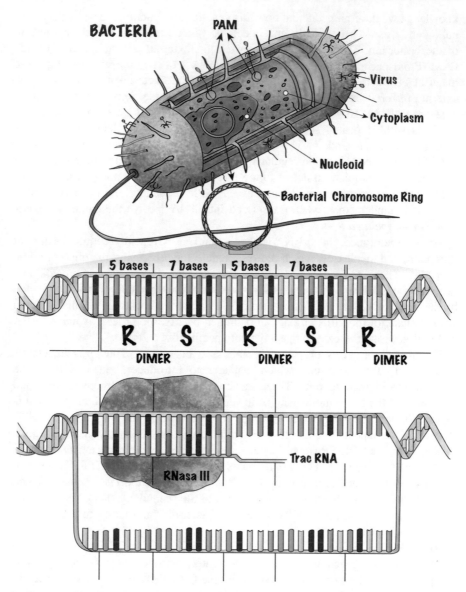

**Fig. 1 Simplification of the mechanism of action of CRISPR/Cas in bacteria**. The virus introduces its genetic material into the bacterial cytoplasm. The cell recognizes a sequence known as the protospacer adjacent motif (PAM) and incorporates the nucleotides adjacent to PAM as a new spacer sequence (S). In this process, repeated sequences (R) assimilated from the virus are duplicated so that spacer sequences are flanked on both sides. Transcription of this newly incorporated material generates a transcript (tracrRNA) complementary to viral repeated sequences that will be further processed to generate a shorten transcript (crRNA). Cas9 will then recruit this crRNA and form the CRIPR/Cas complex. In sum, the crRNA will be responsible for guiding the so-called CRIPR/Cas complex to its target by recognizing complementary sequences when new infections occur. As a consequence, Cas will precisely tear the invading genetic material to pieces

has also been adaptable and applicable to study in a wide range of research fields, such as biotechnology, biomedicine, and personalized gene therapy.

## However, Can the New Molecular "Photoshop" CRISPR/Cas Help in the Cure of Diseases?

So far, so good. We have managed to insert a gene from one organism into another in a precise way, but what is the point of doing this? The "volcanic eruption" over the use of CRISPR/Cas started to grow almost immediately after it was discovered. There have been many recent advances in the manipulation of genomes using this technique, with numerous applications being developed in a very short period of time. This is opening a wide range of uses for CRISPR, from basic research to clinical therapy. Due to the improved precision and accuracy in editing and modifying genomes, it is being studied worldwide in many laboratories dedicated to genome editing, making the CRISPR/Cas system more powerful than most techniques currently used in this field. The result is that many research projects are being accelerated, with highly promising results for therapeutic and biotechnological applications.

The system allows the transcriptional repression or activation of genes involved in alterations, thus controlling the pathological processes that are linked to genetic diseases. This is why the CRISPR/Cas system has been a revolutionary tool. In those pathologies that involve genetic disorders, or those that include genetic risk factors, this genetic "Photoshop" could be used, ideally allowing a pathological nucleotide sequence to revert to a nonpathological sequence.

Research conducted before the discovery of CRISPR paired samples of damaged tissue and healthy tissue from different patients (with the necessary homology to avoid rejection of the tissues). However, it is not easy to perform these pairings in all cases because of the possibility of rejection, or because of the difficulty of access and availability. In light of these challenges, for years, a line of stem cells has been cloned and used. Generated from these clones was a line of healthy cells, which could be transplanted to the sick individual later, and which would restore that part of the damaged tissue. Although this technique proved to be very effective, it was also incredibly expensive and time consuming. The CRISPR/Cas tool has completely changed this process, and allows cloned human cell lines to be generated in a simple and effective way from the same patient, thus reducing the possibility of transplant rejection. With this scientific advance, it has been possible to glimpse the application of this tool in the so-called gene therapies, that is, those treatments aimed at curing a genetic disease, and which are based on the insertion of specific genes in the patient's diseased cells to combat the disease. It has been proven how the use of CRISPR can activate tumor suppressor genes or, on conversely, repress the expression of the so-called oncogenes, which contribute to the development of cancer. In this way, cancerous processes could be controlled; additionally, the use of this genetic tool to *cure* cancer is currently under study, and is the focus of much research work in laboratories around the world.

So far, we have primarily discussed applying the CRISPR/Cas technique in the cure of diseases; it must be said that its use has made it possible to correct the expression of certain altered genes that give rise to congenital disorders. One of these diseases is tyrosinemia, a metabolic modification in the synthesis of tyrosine; for the most part, the symptoms result in damage to the liver and kidney. Another example of a congenital disease is Duchene muscular dystrophy, which is a degenerative muscle disease caused by the alteration of the gene that gives rise to dystrophin, a protein present in muscles. Cystic fibrosis, another congenital disease, causes the accumulation of thick, sticky mucus in different organs of the body, such as the lungs, resulting in chronic and potentially fatal lung disease. In addition, CRISPR/Cas technique can be applied to other congenital diseases including cardiovascular and neurological diseases. All this evidence has confirmed that the CRISPR/Cas genetic tool is an applicable system for the development of gene therapies, hopefully in the near future, whose purpose would be to reverse or treat a disease with a specific genetic alteration.

## Other Applications that Can Benefit from the Use of CRISPR

The use of CRISPR/Cas system in the remodeling of cell DNA has given rise to the possibility of generating new cellular models for the discovery and development of various drugs. In other words, genes that are lethal to cell survival have been identified using cell lines developed by applying the CRISPR/Cas system, giving rise to the testing of new drugs that can be beneficial for cellular organisms. Along this line, genes that confer resistance to certain drugs have also been identified in cells. This has helped to generate new, more appropriate drugs, and has opened a window for the development of new drug therapies less toxic to healthy tissues. Therefore, this genetic system is being used in the pharmacological industry, and it opens a wide range of therapeutic possibilities, as well as potential economic benefits for companies.

Right now, however, we "soldiers in the trenches" care more about the therapeutic possibilities. Those of us with experimental animals, who are dedicated to controlled research with established basic rules, are happy that this revolutionary genetic tool can serve to reduce the number of animals used for research. CRISPR/Cas can be useful in those studies that are aimed at the identification and study of the function of genes in animals, as well as in the identification of genetic mutations associated with biological functional, or even phenotypic (what you see; the appearance), alterations. This tool can be used to recognize genetic mutations in individuals, and can be used to quickly model the genetic variations that have occurred previously, analyze their function, and see the possible repercussions and consequences it may have on the organism. In other words, the use of genetically transformed animals can be done in a simpler, faster way, and with a smaller number of animals used for this purpose, a real achievement and a saving of time, money, and the number of experimental animals used!

In regard to mutant mice, the Photoshopping of the DNA with CRISPR/Cas can generate double and/or triple mutants faster and more economically than can currently be accomplished, without the need to cross strains with each other, and with a smaller number of animals to achieve the desired mutant mouse. The emphasis on the use of fewer animals is a prerequisite for good laboratory practice for research groups working with experimental animals; the use of the number of experimental animals is currently being closely monitored worldwide. Furthermore, it is crucial to understand that initiating a research project and request funding in most countries, requires providing an estimate of the number of animals intended to be used in the research. This estimation is subject to review by the ethical committee and/or animal experimentation center of the institution where the project will be conducted.

How does this genetic tool achieve these benefits? Well, the CRISPR/Cas system can be used directly on the zygote, so it would not be necessary to derive embryos (passing embryos from one litter to another) or perform a coculture with embryonic stem cells, which until now has been one of the limiting steps in the generation of mutants of various species due to ethical issues.

Without the need to generate mutant mice, the advance in the study of CRISPR/Cas has made it possible to directly insert somatic mutations in a certain tissue of the animal at a specific point in time, and thus be able to study more precisely the onset and development of a genetic disease, as well as to know the resulting tissue response to this type of genetic alteration. If all the consequences derived from the application of this genetic tool on the study tissue and on the organism in general could be known, CRISPR could be used to generate mutations that protect tissues from nongenetic diseases. This capacity would be very important to reduce the development of cancer, since the interactions between tumor cells and the rest of the body's cells, especially immune cells, have a drastic effect on the outcome of this disease.

Last but not least, the potential applicability on biotechnology is of great impact. Biotechnology is understood as anything that uses biological systems and living organisms or their derivatives for the creation or modification of biological products or processes for specific uses. In this sense, the applicability of CRISPR/Cas serves to implement more robust metabolic pathways in the industrial biotechnology, and to help in the generation or improvement of industrial products to increase the yield of the production of biofuels or biomaterials, for example.

With respect to agriculture, the CRISPR/Cas system can also lead to improvements in the characteristics of various crops, as well as control their proliferation, so that fruits, vegetables, cereals, and other crops, could be planted outside their natural planting season and have good quality later. In other words, improved food production can feed all the people and animals populating the Earth.

In short, it can be said that the use of this revolutionary molecular tool can improve the lives not only of people but also of animals, plants, and therefore, of our mother Earth.

Despite the many advantages that this genetic system is currently offering, it is still in the testing period. There are still unresolved questions about how the accessibility of the CRISPR affects DNA, and whether generating double-strand breaks affects the correct (and subsequent generation of) proteins important for cell survival. This problem is the one that most concerns the scientific community about the use of CRISPR/Cas system for genetic modification. It should be noted, however, that numerous studies are currently being carried out to resolve this issue, and to understand how to use this tool more precisely in molecular biology.

## CRISPR/Cas and Today's Ethical Reality

The CRISPR/Cas gene editing system, like other systems capable of modifying the genome of different cell types in a targeted and specific way, is not without ethical controversy. The controversy lies in the concern people have when the genomes of human embryos can be modified; it is feared that the application of these genetic methods could be used irresponsibly, creating embryos "à la carte" with specific phenotypic characteristics determined beforehand. This has been the subject of discussion for several years, and has long been regulated by intergovernmental organizations. However, when a new case of genetic manipulation of human embryos appears, even if it is to study the genetic alteration that causes a certain disease, the debate is reopened.

The impressive advances being made in genetic engineering and molecular biology are changing our society and are developing so fast that there is no time for in-depth study or ethical evaluation of the advantages, risks, or even their impact on future generations and the environment. Therefore, there will have been a multitude of scientific publications on new developments from the time these words were written until you are reading them right now. In addition, being a new genetic tool that has not been fully refined, the scientific community is also concerned about its application in patients with genetic diseases, as it could generate mutations in places that are not its target, which could directly affect the patient; if it is a heritable mutation that is even more worrying. Therefore, the transparency of scientists in the results and the research itself, as well as the participation of the public in scientific policies, is of vital importance to achieve progress in genetic engineering.

Despite all the advances using this gene editing tool implies, the CRISPR/Cas system still has a long way to go in optimizing its results; the future of this genomic technology is being written day by day as numerous laboratories all over the world research its operation and its applicability in science, health, and industry. This molecular revolution has taken only its first steps, but it is thought likely that this type of DNA Photoshopping can help to provide feasible solutions to major human problems.

# Bibliography

de la Fuente-Núñez C, Lu TK (2017) CRISPR-Cas9 technology: applications in genome engineering, development of sequence-specific antimicrobials, and future prospects. Integr Biol 9(2):109–122

Hsu PD, Lander ES, Zhang F (2014) Development and applications of CRISPR-Cas9 for genome engineering. Cell 157(6):1262–1278

Lander ES (2016) The heroes of CRISPR. Cell 164(1–2):18–28

Lino CA, Harper JC, Carney JP, Timlin JA (2018) Delivering CRISPR: a review of the challenges and approaches. Drug Deliv 25(1):1234–1257

Lammoglia-Cobo MF, Lozano-Reyes R, García-Sandoval CD, Avilez-Bahena CM, Trejo-Reveles V, Muñoz-Soto RB, López-Camacho C (2016) The revolution in genetic engineering: CRISPR/Cas system. Disabili Res 5(2):116–128

Fernández A, Josa S, Montoliu L (2017) A history of genome editing in mammals. Mamm Genome 28(7–8):237–246

Mojica FJM, Montoliu L (2016) On the origin of CRISPR-Cas technology: from prokaryotes to mammals. Trends Microbiol 24(10):811–820

Montoliu L (2019) Editando genes: recorta, pega y colorea: Las maravillosas herramientas CRISPR. Next Door Publishers S.L.; n°1

**Patricia Corrales Cordón** has a degree in Biology and a master's degree in Molecular and Cellular Biology from the Universidad Autónoma de Madrid (UAM), and obtained her Ph.D. from the Doctoral Program in *Health Sciences* at Universidad Rey Juan Carlos (URJC). She is currently Assistant Professor and Researcher in the Area of Biochemistry and Molecular Genetics of the Department of Basic Health Sciences at Universidad Rey Juan Carlos. She makes part of the High Performance Research Group in the Study of the Molecular Mechanisms of Glycolipotoxicity and Insulin Resistance: Implications in Obesity, Diabetes and Metabolic Syndrome (LipoBeta) and of the Consolidated Research Group in Obesity and Type 2 Diabetes: Adipose Tissue Biology (BIO-FAT) of Universidad Rey Juan Carlos. She is a member of the Spanish Society of Biochemistry and Molecular Biology (SEBBM), the Spanish Society of Endocrinology and Nutrition (SEEN), the Spanish Society of Obesity (SEEDO) and the European Society for the Study of Diabetes (EASD). Her research work focuses on the study of metabolic diseases such as obesity and diabetes, particularly in the modulation of adipose tissue both in animal models and in humans. As a disseminator, she has experience as a speaker at science festivals (e.g., Pint of Science) and outreach activities (e.g., *Science Week*, *Researchers' Night* and the SEBBM *Teacher's Corner*).

**Fernando Muñoz Muñoz** holds a degree in Biology from Universidad Autónoma de Madrid (UAM), a diploma in Environmental Education from the Spanish Association of Environmental Education, a master's degree in Teacher Training for Compulsory Secondary Education and Baccalaureate, Vocational Training and Language Teaching from Universidad de La Rioja (UNIR), and a master's degree in Digital Skills and Computational Processing from Universidad Rey Juan Carlos (URJC). He has developed his career as a teacher at different educational levels: Biology teacher in middle and high school, as well as in the Higher Level Cycles of Clinical and Biomedical Diagnostic Laboratory and Higher Level of Pathological Anatomy and Cytodiagnosis.

# Spinach for Crystal Bones: Popeye Children

Jair Antonio Tenorio Castaño

Popeye the Sailor is a fictional cartoon character who is known for his distinctive appearance, with a pipe in his mouth, a squinty eye, and his trademark sailor outfit. He gains superhuman strength by consuming spinach, which he uses to protect his friends and loved ones. The popularity of Popeye helped boost spinach sales and he was frequently used as a role model for healthy eating.

Who has not dreamed of having super strength like Popeye, being super athletic as Wonder Woman, or being able to fly like Superman, or to climb walls like Spiderman? What if we could eat something like spinach and gain supernatural abilities? Let's keep that in mind as we will be back to this point later.

We have also heard of children with *glass or crystal bones*; what comes to mind is probably the image of a child of short stature, with many broken bones, psychomotor problems, who has difficulty walking, and is overall quite disabled. Sounds like the exact opposite of a superhero, doesn't it? In this chapter we will help you see these children in a different light: as the superheroes they really are. To do that, we are going to explain a bone and metabolic disease that is very little known: hypophosphatasia (HPP).

The occurrence of hypophosphatasia is very low. In Europe, it is estimated that approximately 1 in 300,000 suffers from this disease. Now imagine small, autochthonous, isolated communities, whose reproduction patterns are limited;

J. A. Tenorio Castaño (✉)
Centro de Investigación Biomédica en Red de Enfermedades Raras (CIBERER), Madrid, Spain
e-mail: jaira.tenorio@salud.madrid.org

Institute of Medical and Molecular Genetics, INGEMM-Idipaz, Madrid, Spain

ITHACA, European Reference Network, Brussels, Belgium

Madrid Technology Park (PTM), BITGENETIC, Tres Cantos, Madrid, Spain

© The Author(s), under exclusive license to Springer Nature Switzerland AG 2024     181
M. M. Garcia (ed.), *Tales of Discovery*, https://doi.org/10.1007/978-3-031-47620-4_15

these tend to be more inbred populations. The Canadian Mennonite community is one such population, probably due to a *founder effect* (a common ancestor who transmitted the disease to all his offspring). The frequency of the disease in this community is 1 in 100,000 births. We are talking here about the most severe forms. There are other milder forms, probably underdiagnosed, with a frequency of occurrence much higher than expected—even 1 in 6000 births. That would give us the not negligible figure of approximately 75,000 cases of patients with hypophosphatasia in the European Union and 1.35 million cases worldwide.

The word hypophosphatasia probably leaves a large part of the population indifferent. It reflects the main effect of this disease. The prefix *hypo* comes from the Greek meaning low, and *phosphatasia* (or *alkaline phosphatase* as it is usually called) is a protein in our body essential for our bones to be healthy and above all strong. Individuals with hypophosphatasia have very low or no levels of this protein in the body. You may be wondering what causes this disease, why it affects so few people in the world, or if it has a cure. To answer these questions, we must have to look at the gene (*ALPL*) that encoded the alkaline phosphatase; its genetic pathogenic variants produce a nonfunctional protein, which cannot accumulate in the bones and which causes the mineralization of the bones to be very low (Fig. 1). Imagine that bones are made up of thousands of crystals packed tightly together on a solid coating. The more crystals we have in our bones, the more solid they are and the less likely they are to fracture. This is what synthesized alkaline phosphatase does in our body.

The pathogenic variants that we find in the alkaline phosphatase gene are usually small—just one or a few nucleotides—but that is enough to cause the synthesized protein to act differently than it does in healthy individuals. Pathogenic variants in which there is a loss of a large amount of genetic material (imagine to loss hundreds or thousands letter from your DNA code) are much less common, and account for only 1.3% of cases. If you are wondering what function this protein has, and why it is so important to our bones, read on.

## The Crystal Armor of Bones

Alkaline phosphatase is a protein that is part of a family of enzymes of the same name (*alkaline phosphatases*), some of which are expressed solely in specific tissues (such as in the intestine or placenta). One of them, however, is expressed in many tissues throughout the body: the protagonist of our story, *tissue non-specific alkaline phosphatase*. For this protein to be functionally active, it needs two identical monomers (two single molecules of alkaline phosphatase) to bind together, forming what is known as a *homodimer*. The main function of this activated protein is to *cut* compounds that possess phosphate groups such as *inorganic pyrophosphate* (PPi), *pyridoxal-5-phosphate* (PLP) and *phosphoethanolamine* (PEA). You may not be familiar with these compounds, but they are fundamental in processes in our bodies, such as the transport of vitamin B6 to the brain, or the formation

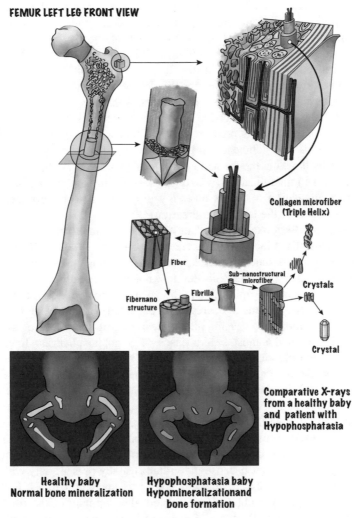

**FEMUR LEFT LEG FRONT VIEW**

Collagen microfiber
(Triple Helix)

Fiber

Sub-nanostructural
microfiber

Crystals

Fibrilla

Fibernano
structure

Crystal

Comparative X-rays
from a healthy baby
and patient with
Hypophosphatasia

**Healthy baby**
**Normal bone mineralization**

**Hypophosphatasia baby**
**Hypomineralizationand**
**bone formation**

**Fig. 1  Diagram of bone formation and components.** Details of the participation of hydroxyap-atite bone crystals in the formation of the femur from a complete longitudinal view, breaking down each of its components. In hypophosphatasia, these crystals are altered due to the high concentra-tion of inorganic phosphorus (PPi) in the extracellular bone matrix. This inhibits the nucleation of calcium and phosphate crystals, which in turn interferes with the formation of hydroxyapatite crys-tals. Since hydroxyapatite crystals participate in the formation of collagen fibers in bones, which will later lead to the formation of the final structure of the bone with its individual trabeculae, the consequence of abnormal production leads to skeletal alterations. The image below shows the femur of a healthy person and the femur of a patient with hypophosphatasia. In the latter, a reduc-tion can be seen in the formation of bone, both in its mineralization and its structure, due to the lack of the enzyme tissue non-specific alkaline phosphatase (TNSALP). This deficiency is caused by mutations in the gene that encodes the enzyme, ALPL

of the hydroxyapatite crystals that accumulate in the bones for their mineralization and to prevent them from breaking (Fig. 2). During embryonic development, alkaline phosphatase participates in the formation of the rib cage; Ribs need to be mineralized to function properly. When the rib cage does not develop properly due to a malfunction of the protein, the lungs are not able to expand properly, causing respiratory problems. Therefore, the deficiency of alkaline phosphatase is related to the accumulation of these phosphorus substrates (PPi, PLP and PEA) at the extracellular level (i.e., outside the cells), which will result in the clinical manifestations and symptoms presented by patients with hypophosphatasia. This accumulation leads to a deficiency in bone mineralization because PPi within its functions acts by blocking the formation of hydroxyapatite crystals, which are fundamental for bone mineralization (which gives us healthy and strong bones). This is why the accumulation of PPi is inversely proportional to bone mineralization (the more PPi there is, the less mineralization).

Remember that Pyridoxal-5-phosphate (PLP) is the main form of vitamin B6 transported across the blood–brain barrier to the brain. Vitamin B6 is naturally present in many foods (such as chickpeas, tuna, salmon, potatoes, etc.) and is involved in more than a hundred biochemical reactions, including the metabolism of amino acids, sugars and lipids. Thus, the accumulation of PLP due to alkaline phosphatase deficiency results in decreased dephosphorylation of PLP to pyridoxal phosphate (or pyridoxine), causing the appearance of neurological symptoms, including seizures in the most severe cases of hypophosphatasia. Interestingly, some patients with pathogenic variants in the alkaline phosphatase gene do not completely lose the functionality of the protein; residual activity can be detected, which means that, this protein can act (although at half speed), and therefore, the symptoms in these patients are usually less severe.

It is important to note that hypophosphatasia can appear throughout one's lifetime. The most severe forms, however, usually appear in childhood, and the mildest forms in the adult period. In fact, in these mild forms in adults it is very difficult to establish a clinical diagnosis because individuals usually present only with premature loss of teeth and/or muscle pain, which are very specific features. In the more severe forms, respiratory problems (due to poor development of the rib cage), or skeletal problems (such as fractures of the long bones) are usually observed. The clinical range in this disease is very variable.

## Personalized Medicine to the Rescue

You may be wondering how we can help these patients to lead a better life within their disease. Are there treatments? Is there a cure? Thanks to advances in recent years, several clinical trials have demonstrated the efficacy of a specific enzyme

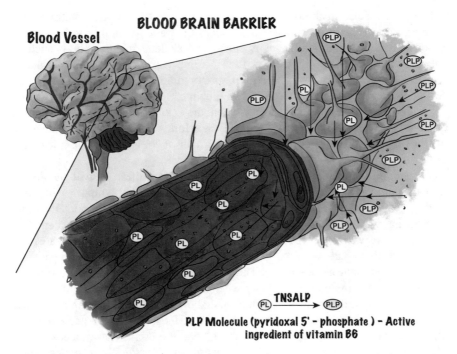

**BLOOD BRAIN BARRIER**

**Blood Vessel**

**TNSALP**

**PLP Molecule (pyridoxal 5' – phosphate ) – Active ingredient of vitamin B6**

**Fig. 2 Transport of vitamin B6 to the brain**. Vitamin B6 plays a fundamental role in the brain. It participates in the synthesis of some neurotransmitters, such as dopamine, serotonin, γ-aminobutyric acid (GABA), noradrenaline, and the hormone melatonin. The synthesis of these neurotransmitters is differentially sensitive to the levels of vitamin B6, which is considered a very water-soluble molecule unable to cross the blood–brain barrier. As a general rule, only small, highly fat-soluble (not very water-soluble) compounds can access the brain. Vitamin B6 therefore enters the brain in the form of a precursor, pyridoxal-5-phosphate (PLP). Once in the central nervous system, PLP can constitute vitamin B6, which, when degraded after playing its role as a coenzyme in different cellular (metabolic) processes, again generates PLP (and amino acids). As PLP is an active and therefore reactionary molecule, it is converted by the action of alkaline phosphatases into pyridoxal (PL) and can accumulate as a reservoir in the cell. Other cellular enzymes, such as phosphate kinases, participate in the continuous recycling cycle of vitamin B6, converting PL into PLP. Therefore, a deficiency of alkaline phosphatase will lead to an accumulation of PLP, and it will not be able to carry out its functions correctly; this correlates with the clinical manifestations at the level of the central nervous system that we find in people who present hypophosphatasia

replacement therapy using a *recombinant enzyme*[1] directed specifically at bone tissue (ERT) *(StrensiqTM-asfotase alfa)*, which has very good results in patients with hypophosphatasia. Here is how it works: Imagine that these children are given a drug that contains the alkaline phosphatase that their own organism cannot synthesize or do it in a wrong way, and that drug is capable of restoring function of

---

[1] Recombinant enzymes are chimeric or heterologous proteins obtained by expressing a cloned gene in a species or cell line other than the original cell, such as human insulin generated in vitro.

alkaline phosphatase in the different organs in which it acts. This treatment has shown an improvement in the survival rate and the patients' ability to mineralize the bones, reducing the number of fractures and skeletal manifestations such as joint pain or the underdevelopment of the rib cage. It has even been seen that, in just a few months, a child was able to almost completely regain the ability walk after being treated with this recombinant enzyme. Imagine their reaction when their body responded as they wanted, when they could walk normally (almost run!), and with hardly any fatigue, because their lungs also worked better. Surely it was a very similar reaction to the one Superman or Wonder Woman must have had the first time they flew. Wow!

The medicine contains the enzyme that is missing on these individuals due to a genetic pathogenic variant. In a relatively short time, the symptoms improve and they have a better quality of life. While some adverse effects related to this treatment have been described, these are not very serious and the frequency of occurrence is very low. So, can we treat children with this drug, and is it safe? All the studies done to date show the high efficacy of this treatment in these patients. Making the treatment more specific (i.e., adding the enzyme in a vehicle that can transport it only to the part of the body where it is needed) means that it does not act in other parts of the body, thus reducing side effects. This is the "spinach" that our "Popeye children" need: an advanced genomic technology that starts with the simple idea that we have always been told to "give the body what it needs."

In recent years, you have probably heard of "personalized or precision medicine," and you may be wondering what that means, what it is for, and whether there was no personalized medicine before. It is the selection and treatment of individuals, taking into account their specific clinical profile. It is also known as genomic medicine. The first step, then, in treating a patient with hypophosphatasia is to determine whether there is a genetic alteration in their alkaline phosphatase gene. Obvious, right? For this purpose, genetic techniques known as *next-generation sequencing* (NGS), was applied. Once each patient's candidate genetic variants are obtained, they are classified according to whether they caused the disease. Through a series of bioinformatics processes, once the variant that causes the disease has been detected we can diagnose the patient and offer treatment with our recombinant enzyme. It sounds very simple, doesn't it? In genomic medicine or personalized medicine, simplicity and the simplest ideas are usually the ones that work best.

## Different Country's Genetic Background Implies Different (Pathogenic) Gene Variants

Today, we know that the human genome carries variations between populations from different regions or countries. It is crucial for genetic studies of a country's population to be both precise and comparable due to the distinct nature of the gene pool among different populations. Interestingly, there are variants unique to just one population or even to one small human group. This means that there are genetic pathogenic variants that cause hypophosphatasia and that are specific only to one region's population. Remember that this phenomenon is referred to as the founder effect—a variant that originated, got transmitted across generations, and persisted within a particular population.

As we know, proteins are made up of many links, and each of these links is an amino acid. We can estimate the importance of these amino acids according to the conservation along the orthologues of the protein in other species. The more conserved an amino acid is, the more important it is for the function of the protein. If we look at the orthologues of this protein in other species, we see that the amino acids where the mutations that cause hypophosphatasia are found are overwhelmingly highly conserved throughout evolution—even as far back in species such as the fruit fly or *Drosophila melanogaster*. This suggests evidence for the importance of this amino acid in the function of the enzyme alkaline phosphatase.

Undoubtedly, the real heroes of this story are our Popeye children. Those who, despite having a disease, are able to move forward and, with the help of scientific advances, improve their quality of life. There was the case of a family with three siblings, all of whom had hypophosphatasia. You may think "how can this be?" Or perhaps you will wonder if it simply bad luck. Looking deeper into the family history, the parents of these three siblings were related, which means that there were inbreeded. Consanguinity increases the risk of having offspring with autosomal recessive diseases, such as severe hypophosphatasia. In another family, we found a five-year-old girl affected with a homozygous pathogenic variant; that is, the two copies she had of the gene were not functional. The girl had delayed closure of the anterior fontanel, moderate to severe skeletal problems, short stature, delayed dental eruption, recurrent healed fractures since the age of three with callus, bowed long bones, and hollowing of the metaphyses. She is the daughter of a consanguineous couple who had another child with severe hypophosphatasia (whose genetic study could not be performed). This was the first time this variant was detected in homozygosis. This variant has been described by other groups. The girl's parents carried this mutation in heterozygosis but, surprisingly, did not manifest any clinical features associated with hypophosphatasia. This only highlights the high clinical variability of hypophosphatasia, and the possible residual functional effect of alkaline phosphatase.

Infantile hypophosphatasia was evaluated in patients who had their first clinical hypophosphatasia-related episode after birth and before six months of life. In general, the inheritance pattern was autosomal recessive, as in the perinatal group, but there were some cases with an autosomal dominant pattern. As suggested by previous authors, autosomal dominant pathogenic variants in the alkaline phosphatase gene are usually associated with childhood, adolescence and adulthood age of onset because the other normal allele provides residual enzyme function sufficient to reduce the severity of skeletal manifestations. Infantile hypophosphatasia is considered when the first clinical manifestations appear between 6 months and 18 years of age and are mainly characterized by rickets, muscle weakness, bone pain, poor physical performance, and dental abnormalities. Infantile hypophosphatasia can be inherited in an autosomal recessive or dominant pattern. In contrast, adult hypophosphatasia is classified according to individuals in whom clinical manifestations appear after the age of 18 years. Despite this classification, many adults reported having symptoms of hypophosphatasia during childhood, but were not diagnosed until later in life. As you may know, one of the mildest manifestations of this disease is premature tooth loss, which could be a symptom that often goes unnoticed or underdiagnosed. Although the possibility that premature tooth loss is ultimately hypophosphatasia is very low, it is important to emphasize that it is a possibility. Individuals with high-impact mutations in the alkaline phosphatase gene have been identified who had only these dental problems. *High-impact mutations* are so-called because they result in the synthesis of a prematurely truncated protein, smaller than a healthy native protein. It can even result in the erroneous protein not being formed.

## Spinach for Everyone

In many cases of patients in whom hypophosphatasia is suspected, the genetic study has not confirmed a pathogenic variant; therefore, we wondered if these patients truly have hypophosphatasia, and if not, what do these patients have? It is possible that a differential diagnosis for some of these patients may include other skeletal dysplasia with similar clinical characteristics, such as osteogenesis imperfecta, or even others even less frequently seen, such as campomelic dysplasia, hypophosphatemic rickets, cleidocranial dysplasia, or osteoporosis. What is clear is that giving our "spinach" to these patients, who do not actually have hypophosphatasia, will not make them better.

Many questions remain regarding hypophosphatasia. Is it possible that there are other genes involved? Can the environment have an effect on the clinical manifestations of the disease? It is obvious that more studies still need to be performed.

Treatment with the enzyme replacement commercially known as *asfotase alfa*, has been shown to improve the symptoms of patients diagnosed with hypophosphatasia. The symptoms of a 16-year-old boy who has been treated for a year with the recombinant enzyme have substantially improved. In three months of treatment, the chronic pain he suffered daily, and which required he take painkillers, disappeared. Additionally, he was able to again walk without need of a cane (for him, that might be pretty close to the sensation of flying). This improved his quality of life and reduced visits to specialists (because who likes going to the doctor?) with all that entails, including hospital admissions and additional medications. When replacement therapy was discontinued, recurrence of bone demineralization was observed. Some clinical manifestations, such as craniosynostosis and nephrocalcinosis, persist even after treatment due to the natural course of this disease. This motivates us to continue working on new, more effective therapies to improve all the patients' symptoms—until the disease can be cured. The adverse effects reported and associated with enzyme replacement therapy include skin reactions at the injection site (such as erythema, pain, and induration), with a frequency of occurrence of approximately 75% of patients treated.

When we think about diseases, we often focus only on the negative aspects; but as we have seen throughout this chapter, the stories can be told in other ways. Hypophosphatasia is an infrequent disease that varies in severity, but today, there is an effective treatment; not all the symptoms are improved, but the most severe ones are. People suffering from this disease have hope of an improved quality of life—of being able to do things that we take for granted: walking, running, or play freely. The aim of the new drugs in personalized medicine, or pharmacogenomics, is to give everyone what they need on an individualized basis. We hope this chapter contributes to a better understanding of the clinical characteristics of patients who have this disease, and provides extremely valuable information for correct genetic counselling and clinical follow-up of patients. Understanding and information are important not only for the clinical genetics unit, but also for all specialists who treat patients with hypophosphatasia (rheumatologists, pediatricians, neonatologists, surgeons, endocrinologists, radiologists, dentists), and for all patients with hypophosphatasia. The best of this story is, undoubtedly, yet to come.

# Bibliography

Berger J, Garattini E, Hua JC, Udenfriend S (1987) Cloning and sequencing of human intestinal alkaline phosphatase cDNA. Proc Natl Acad Sci USA 84(3):695–698

Bowden SA, Foster BL (2018) Profile of asfotase alfa in the treatment of hypophosphatasia: design, development, and place in therapy. Drug Des Devel Ther 12:3147–3161

Drake MT, Khosla S (2008) Bone-targeted replacement therapy for hypophosphatasia. J Bone Miner Res Off J Am Soc Bone Miner Res 23(6):775–776

Greenberg CR, Taylor CL, Haworth JC, Seargeant LE, Philipps S, Triggs-Raine B et al (1993) A homoallelic Gly317->Asp mutation in ALPL causes the perinatal (lethal) form of hypophosphatasia in Canadian Mennonites. Genomics 17(1):215–217

Henthorn PS, Raducha M, Fedde KN, Lafferty MA, Whyte MP (1992) Different missense mutations at the tissue-nonspecific alkaline phosphatase gene locus in autosomal recessively inherited forms of mild and severe hypophosphatasia. Proc Natl Acad Sci USA 89(20):9924–9928

Hu JC, Plaetke R, Mornet E, Zhang C, Sun X, Thomas HF et al (2000) Characterization of a family with dominant hypophosphatasia. Eur J Oral Sci 108(3):189–194

Hypophosphatasia R-G (2013) Pediatric endocrinology reviews. PER 10(Suppl 2):380–388

Kam W, Clauser E, Kim YS, Kan YW, Rutter WJ (1985) Cloning, sequencing, and chromosomal localization of human term placental alkaline phosphatase cDNA. Proc Natl Acad Sci USA 82(24):8715–8719

Leung EC, Mhanni AA, Reed M, Whyte MP, Landy H, Greenberg CR (2013) Outcome of perinatal hypophosphatasia in Manitoba Mennonites: a retrospective cohort analysis. JIMD Rep 11:73–78

Lia-Baldini AS, Muller F, Taillandier A, Gibrat JF, Mouchard M, Robin B et al (2001) A molecular approach to dominance in hypophosphatasia. Hum Genet 109(1):99–108

Macfarlane JD, Poorthuis BJ, van de Kamp JJ, Russell RG, Caswell AM (1988) Hypophosphatasia: biochemical screening of a Dutch kindred and evidence that urinary excretion of inorganic pyrophosphate is a marker for the disease. Clin Chem 34(9):1937–1941

Macfarlane JD, Poorthuis BJ, Mulivor RA, Caswell AM (1991) Raised urinary excretion of inorganic pyrophosphate in asymptomatic members of a hypophosphatasia kindred. Clinica chimica acta; Int J Clin Chem 202(3):141–148

Millan JL, Manes T (1988) Seminoma-derived Nagao isozyme is encoded by a germ-cell alkaline phosphatase gene. Proc Natl Acad Sci USA 85(9):3024–3028

Millan JL, Narisawa S, Lemire I, Loisel TP, Boileau G, Leonard P et al (2008) Enzyme replacement therapy for murine hypophosphatasia. J Bone Miner Res Offi J Am Soc Bone Miner Res 23(6):777–787

Mornet E (2007) Hypophosphatasia. Orphanet J Rare Dis 2:40

Mornet E, Yvard A, Taillandier A, Fauvert D, Simon-Bouy B (2011) A molecular-based estimation of the prevalence of hypophosphatasia in the European population. Ann Hum Genet 75(3):439–445

Mornet E (2015) Molecular genetics of hypophosphatasia and phenotype-genotype correlations. Subcell Biochem 76:25–43

Moulin P, Vaysse F, Bieth E, Mornet E, Gennero I, Dalicieux-Laurencin S et al (2009) Hypophosphatasia may lead to bone fragility: do not miss it. Eur J Pediatr 168(7):783–788

Rougier H, Desrumaux A, Bouchon N, Wroblewski I, Pin I, Nugues F et al (2018) Enzyme-replacement therapy in perinatal hypophosphatasia: case report and review of the literature. Arch Pediatr 25(7):442–447

Russell RG (1965) Excretion of Inorganic Pyrophosphate in Hypophosphatasia. Lancet 2(7410):461–464

Simon S, Resch H, Klaushofer K, Roschger P, Zwerina J, Kocijan R (2018) Hypophosphatasia: from diagnosis to treatment. Curr Rheumatol Rep 20(11):69

Tenorio J, Alvarez I, Riancho-Zarrabeitia L, Martos-Moreno GA, Mandrile G, de la Flor CM et al (2017) Molecular and clinical analysis of ALPL in a cohort of patients with suspicion of Hypophosphatasia. Am J Med Genet A 173(3):601–610

Ucakturk SA, Elmaogullari S, Unal S, Gonulal D, Mengen E (2018) Enzyme replacement therapy in hypophosphatasia. J Coll Physicians Surg Pak 28(9):S198–S200

Whyte MP, Greenberg CR, Salman NJ, Bober MB, McAlister WH, Wenkert D et al (2012) Enzyme-replacement therapy in life-threatening hypophosphatasia. N Engl J Med 366(10):904–913

Whyte MP, Zhang F, Wenkert D, McAlister WH, Mack KE, Benigno MC et al (2015) Hypophosphatasia: validation and expansion of the clinical nosology for children from 25 years experience with 173 pediatric patients. Bone 75:229–239

Zurutuza L, Muller F, Gibrat JF, Taillandier A, Simon-Bouy B, Serre JL et al (1999) Correlations of genotype and phenotype in hypophosphatasia. Hum Mol Genet 8(6):1039–1046

**Jair Antonio Tenorio Castaño** holds a degree in Biology, a master's degree in Pharmacological Research and a Ph.D. in Biological Sciences from Universidad Autónoma de Madrid-Insituto de Investigación Hospital Universitario La Paz (UAM-IdiPAZ). He has conducted different internships at national and international research centers and is currently a researcher in Hospital Universitario La Paz. He is part of the Center for Biomedical Research Network on Rare Diseases (CIBERER), the Institute of Medical and Molecular Genetics (INGEMM) and is a member of the Foundation against Pulmonary Hypertension (FCHP). His research work focuses on the analysis and diagnosis of human genetic diseases.

# Olive Oil: A Seasoning and Something More

Gerardo Ávila Martín⬤

For centuries the Mediterranean diet, rich in olive oil, has had health benefits for those who have followed it. With the advance of new technologies during the last century, the benefits of olive oil have begun to be investigated and recognized worldwide. Its health effects seem to be due to its composition, rich in fatty acids, particularly oleic acid. Epidemiological studies suggest that olive oil has many benefits: it has a protective effect against certain types of cancer (breast, prostate, endometrial, and colon); it may be partly responsible for the low incidence of certain cardiovascular diseases, and helps to keep blood pressure at normal and stable levels; it regulates cholesterol due to the antioxidant effects of its components. Daily consumption of olive oil enhances cognitive function, prevents stroke, reduces the risk of diabetes and obesity, and prevents osteoporosis and rheumatoid arthritis due to its anti-inflammatory properties. In addition to its healing properties, olive oil regenerates skin cells and softens the epidermis due to its antioxidant and nourishing properties. It is widely used in rejuvenating masks, it replenishes dry and damaged hair, is used in shaving creams, and in eyelash and nail treatments, and is incredibly popular among lovers of natural remedies. Last but not least, medicinal teas can be made from the leaves of olive trees. Olive leaf tea is the perfect alternative to green tea, as it contains no caffeine. So, are we talking about a superfood or a "medicine" with a potential yet to be discovered?

G. Ávila Martín (✉)
Health Integrated Area, Research Support Unit, Hospital General Universitario Nuestra Señora del Prado, Servicio de Salud de Castilla-La Mancha (SESCAM), Talavera de la Reina, Toledo, Spain
e-mail: gavila@sescam.jccm.es

M. M. Garcia (ed.), *Tales of Discovery*, https://doi.org/10.1007/978-3-031-47620-4_16

## Grandma Was Right

For years, olive oil has been used to prepare home remedies for various pathologies and illnesses (Fig. 1). Here is a partial list of these home recipes:

| | |
|---|---|
| Arthrosis creams | Mixture of virgin olive oil with 100 g of dried chamomile flowers applied to the joints |
| Tobacco addiction | Ingestion of five teaspoons of oil on an empty stomach |
| Acne | Mixture of 100 drops of lavender essence oil and ¼ l of oil applied to the affected area |
| Gingival health | Chewing olive leaves |
| Constipation | Chamomile infusion with a couple of teaspoons of virgin olive oil |
| Wound healing | Dressing with olive oil by mixing a rue, a dried and powdered wild herb, to plug the wound |
| Preventing gallstones | Squeezed lemon juice mixed with a tablespoon of olive oil taken on an empty stomach |
| Eczema | Topical application of olive oil |

Why, however, are we talking about homemade recipes when we could be talking about scientific protocols? In this chapter, we will pay special attention to the anti-inflammatory "superpowers" that olive oil possesses. Inflammation is a natural process, similar to pain, it protects us on the vast majority of occasions. It is usually accompanied by a sensation of heat due to the swelling of the tissues affected by the accumulation of liquids. Inflammation also produces pain. In acute processes, when you get a contusion, for example, the inflammation lasts from a

**FROM TRADITION TO CLINICAL APPLICATION**

**Fig. 1 From traditional to clinical application.** Medicinal preparations based on olive oil

few minutes to a few days, and it is not always necessary to resort to an anti-inflammatory to relieve it. However, when the situation is not transient, the body needs to mitigate the effects of inflammation, and anti-inflammatory treatments are proposed. Simply put, anti-inflammatory treatments help reduce inflammation. There are many remedies that have long been used for this purpose:

| Prevention/decrease of body inflammation | Consume a moderate dose of olive oil daily |
|---|---|
| Otitis | Pour a couple of drops of oil into the ear, and clean with a cotton swab |
| Hemorrhoids | Soak a gauze pad with a little oil and soda, and apply to the area until it shrinks |
| Sunburn | Apply a mixture of oil and egg whites to soothe discomfort |
| Trauma | Pour a few drops of oil on a cotton pad previously rubbed with a piece of garlic, and apply on the area to reduce swelling |

As you can see, many home remedies have been used for years, but it is important to know if the beneficial properties surrounding this potential medicine are knowledge-based. Are these remedies myths or, on the contrary, do they have a scientific basis? In the following section, we will try to shed some light on these questions (Fig. 1).

## Science Explains It

As we start exploring of this field, we can find that there is scientific literature that demonstrates the effects of the main *active ingredients*[1] of olive oil. The composition of olive oil varies according to its place of origin, the variety of olive used, and the quality of the oil. It is made up of a mixture of approximately 99% fatty acids, of which, depending on the variety of olive, 62–82% is represented by oleic acid, its most important active ingredient, and implicated in many of the beneficial effects of olive oil. Fatty acids are biological molecules that are formed from a very long, linear chain of hydrogens and carbons. To understand what fatty acids truly are, it is necessary to have some specific knowledge about them. When we speak of fatty acids, we often use the colloquial term "fats," but in this chapter we use the term "lipids."

The types of fatty acids that we eat are important because they can cause changes in the membranes of our cells, and can influence the transmission of signals from these cells to the rest of the body; this can lead to the development, or reversal, of disease. The process of transmitting signals from one cell to another

---

[1] The concept of *active ingredient* is used in chemistry to name the compound that carries the pharmacological qualities present in a substance. This means that the active ingredient of a medicine is the one that prevents, treats or cures a health disorder or disease.

called *signal transduction*. It is a process by which our cells respond to substances in the environment around us. The binding of a substance to a molecule on the cell membrane causes signals to pass from one molecule to another within the cell. These signals can affect both the cell's multiplication and its death. The main characteristic of the cell membrane is its selective permeability, which allows it to select the molecules that should enter and leave the cell. If we modify these lipids, could scientists control the selective permeability of the plasma membrane? Could the intake of specific molecules dedicated to regulating membrane lipids levels prevent and reverse pathological processes? We will try to answer these questions.

Everything we ingest in our diet produces effects in our organism. In particular, the fatty acids we eat regulate the lipid composition of the membranes of our cells. These fatty acids can be *saturated* and *unsaturated*. The definition of saturated and unsaturated fats is complex, and requires some knowledge of chemistry to understand it well. In general saturated fats have all their carbon atoms bound together, and are known as "bad fats"; unsaturated fats have one (monounsaturated) or several (polyunsaturated) carbon atoms always free, and are known as "good fats" (Fig. 2). Their different compositions give them either beneficial or detrimental health properties. Numerous studies indicate that consumption of saturated fats is associated with impaired health. Certain vascular and tumor diseases have been directly related to the intake of this type of lipid. In contrast, unsaturated fatty acids have long been linked to the prevention or reversal of certain diseases. Omega 9 series are grouped in the monounsaturated fatty acids, and oleic acid is also included in this family. The term "omega" refers to the number of carbons that appear before the free carbon atom that characterizes these fatty acids. In the case of oleic acid, there are 9 carbons until the first and only free carbon atom appears.

There are multiple scientific studies related to the beneficial properties of monounsaturated fatty acids. Among other things, it has been shown that the replacement of saturated fats in the diet with monounsaturated fats is associated with increased daily physical activity and resting energy expenditure, and a decrease in anger and irritability. Additionally, the levels of oleic acid along with other monounsaturated fatty acids in red blood cell membranes are associated with a protective effect against the risk of breast cancer. The consumption of monounsaturated oils is also associated with a much healthier blood lipid content in children. Furthermore, a Mediterranean diet rich in oleic acid reduces pain in patients affected by inflammatory arthritis. Omega 9 fatty acids are also involved in a wide variety of biological functions; from a scientific point of view, they have some very interesting superproperties to explore.

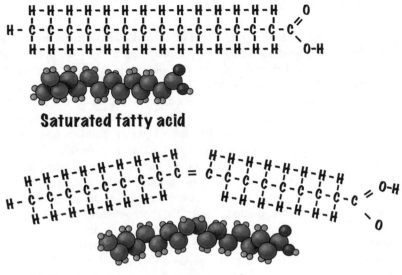

**Saturated fatty acid**

**Monounsaturated fatty acid**

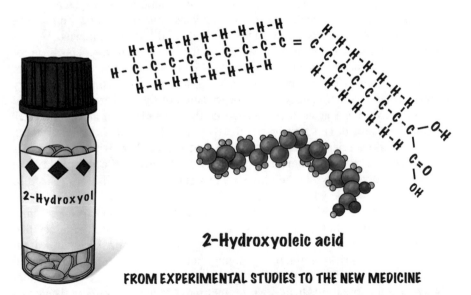

**2-Hydroxyoleic acid**

**FROM EXPERIMENTAL STUDIES TO THE NEW MEDICINE**

**Fig. 2 Structures of saturated and monounsaturated fatty acids.** When fatty acids solidify, their molecules are packed together so that each is bound to its neighbors by *van der Waals* interactions between the respective hydrocarbon chains. Unlike saturated fatty acids, unsaturated fatty acids have one or more double bonds between the carbon atoms of their long hydrocarbon chains. This affects the distance between the atoms of these carbons and the bond angles, forming "bends" that result in fewer interactions with the carbon chains of other neighboring fatty acid molecules. Hence, its melting point is lower, and its properties are different. Oleic acid is an 18-carbon unsaturated fatty acid with a double bond at position 9. In the future, small modifications in oleic acid molecules, such as the 2-hydroxyoleic acid shown in the image, could make it a drug for clinical use

## Do We Have Leads to Develop a New Drug?

There are several scientific studies that affirm oleic acid's involvement in a multitude of cellular mechanisms that take place in our central nervous system. In recent years, several researchers have shown that oleic acid works as a modulator of communication between our neurons. Our neurons communicate with each other through *neurotransmitters* (biological molecules involved in the transmission of information between neurons), to control specific functions of our nervous system. Oleic acid has the ability to block certain neurotransmitters involved in inhibiting or reducing neuronal activity. It can also act as an activator of specific types of receptors, particularly those involved in the activation of *serotonin* (a neurotransmitter involved in the modulation of various biological and neurological processes, such as aggression, anxiety, appetite, cognition, learning, memory, mood, nausea, sleep, and the regulation of our body temperature). We also know that oleic acid has a high affinity for certain receptors inside cells; the receptors are involved in our body's various physiological processes: it specifically activates a subtype of this receptor and regulates the growth of neurons.

It has been postulated for years that our body has a mechanism responsible for the production of oleic acid. Studies have shown that the endogenous synthesis of oleic acid increases in response to an injury to the central nervous system, which highlights its role in the regeneration of the nervous system. What do we make of this discovery? Researchers are presented with a world full of unknowns.

However, back to inflammation. We know that oleic acid has a specific protective effect. It decreases *lipid oxidation* of certain cell types, and shows an anti-inflammatory effect by inhibiting the production of certain inflammatory chemicals. Lipid oxidation refers to the degradation of lipids; therefore, oleic acid has a protective effect against the substances that degrade them and participates in the maintenance of cell membranes that we have described above. Everything is related! Various scientific studies have provided researchers with enough clues to postulate that the development of a drug from "superoleic acid" could have an interesting applications in the clinical treatment for certain diseases related to abnormalities in our nervous system.

This leads us to consider chronic pain—a disabling, difficult-to-treat, sometimes long-lasting pathology which does not always coincide with the injury that originally produced it. People with this pathology report abnormal and unpleasant sensations such as pricking, tingling, stinging, burning, or discomfort. There is no specific treatment to alleviate the effects of chronic pain; at the moment, treatment is based on a mixture of medicines with sometimes adverse side effects. There is a critical need to discover new experimental drugs with significant curative potential, free of the side effects that diminish the patient's quality of life. All the verified and available scientific information regarding the biological effects of oleic acid indicate that further study of its therapeutic potential to generate new treatment for chronic pain has extraordinary potential.

In recent years, animal studies have shown that oleic acid treatment in combination with albumin enhances the recovery of locomotor activity after injury to the central nervous system. Albumin is a protein that exists in our blood plasma, and has specific binding sites for oleic acid. The researchers realized that albumin acts as a natural transporter of oleic acid, facilitating its distribution throughout the body. In addition, this experimental treatment repairs the functioning of the downstream pain inhibition systems, enhancing the presence of substances involved in the repair, such as the previously mentioned serotonin. These descending pain inhibition systems are endogenous systems involved in the control of pain sensation. In situations of injury to the nervous system, these systems are deregulated and do not control the painful sensation; therefore, it may be that the pain lasts indefinitely. They work like a "switch" that is ON to attenuate the pain, and OFF when it is not needed. In situations of chronic pain, the switch is broken and does not turn ON (attenuate the pain) as it should; the painful sensation can last for a long time and even be magnified. In rodents, oleic acid boosts the recovery of this system and seems to help repair the damaged switch. In these experiments, the treatment caused an increase in serotonin below a lesion generated in the central nervous system. Oleic acid functions as a "key" that opens the serotonin receptors, which function as a neurotransmitter involved in the correct functioning of the descending pathways that inhibit pain. The researchers concluded that animals that were treated with oleic acid had an increase in serotonin compared to animals which were not. In the same study, it was also shown that the administration of oleic acid treatment produces a reduction in the activation of receptors related to the transmission of pain information, and a reduction in the activation of *microglial cells* (a central nervous system cell which has recently been shown to be involved in inflammation, development, and chronification of injury-related pain, through specific receptors). This is getting even more interesting!

These mechanisms of action suggest that oleic acid functions as an analgesic and a neuronal growth enhancer in central nervous system injury, which positions oleic acid as a potential drug to be considered for clinical application. In recent years, work has been done on a new modified oleic acid molecule (*2-hydroxyoleic acid*), which undergoes a slower metabolization; therefore, the time it is available in the body is much longer compared to its counterpart oleic acid. This new drug has impressive hypotensive and anticarcinogenic properties. Recently, animal studies have shown that the administration of 2-hydroxyoleic acid significantly reduces pain in response to painful stimuli, and reduces microglial cell activation. It also causes a reduction in anxiety behavior as an effect associated with pain in animals that have lesions of the peripheral nervous system. The latest studies, conducted in 2017, are based on the treatment of animals with the albumin-2-hydroxyoleic acid complex after central nervous system injury. Albumin continued to function as a natural transporter, as in the experiments with unmodified oleic acid. To investigate the specific mechanisms of action, the researchers carried out a gene analysis in the area of nerve injury in the spinal cord. They tested whether experimental treatment with 2-hydroxyoleic acid influenced the coding of molecules related to inflammation and chronic pain. Representative genes involved in pain and nervous system regeneration were selected and analyzed. Comparison of gene expression between animals treated and untreated with 2-hydroxyoleic acid revealed relevant changes in the expression of genes associated with neuronal growth, survival of neurons and genes related to pain and inflammation. It was shown that 2-hydroxyoleic acid treatment causes not only a significant increase in the expression of genes related to cell growth, but also causes significant reductions in genes related to inflammation. Thus, 2-hydroxyoleic acid-albumin was shown to decrease the activation of genes involved in inflammation, and increase the activation of genes involved in neurite outgrowth, balancing the body's important response to central nervous system injury. All these gene changes could corroborate data from previous studies, and position 2-hydroxyoleic acid as a drug that could improve recovery from paralysis and chronic pain that occurs after a nervous system injury. Spectacular!

So many new discoveries! Let's summarize: Science has corroborated the astonishing properties of this supermedicine. Scientists have discovered that treatment with omega 9 monounsaturated fatty acids, both with oleic acid and its synthetic derivative 2-hydroxyoleic acid, either individually or combined with albumin, reduces pain and modulates anxiety associated with pain in peripheral nervous system injuries in animal models. In addition, it has been found that these molecules improve sensory and motor function, and modulate genes involved in physiological recovery after injury to the central nervous system. Therefore, the properties of these fatty acids makes them interesting agents for administration in pathologies associated with nervous system damage. In the long term, it will be necessary to carry out clinical trials to demonstrate its effect, and to improve the lives of patients with chronic pain and associated pathologies.

In this chapter, we have shown an example of how home remedies used by our grandmothers have a powerful scientific basis. They present scientists with fascinating clues for the development of new drugs with extraordinary potential.

# Bibliography

Alberts GL, Chio CL, Im WB (2001) Allosteric modulation of the human 5-HT(7A) receptor by lipidic amphipathic compounds. Mol Pharmacol 60(6):1349–1355

Alemany R, Terés S, Baamonde C, Benet M, Vögler O, Escribá PV (2004) 2-hydroxyoleic acid: a new hypotensive molecule. Hypertension 43(2):249–254

Alemany R, Vögler O, Terés S, Egea C, Baamonde C, Barceló F, Delgado C, Jakobs KH, Escribá PV (2006) Antihypertensive action of 2-hydroxyoleic acid in SHRs via modulation of the protein kinase A pathway and Rho kinase. J Lipid Res 47(8):1762–1770

Alemany R, Perona JS, Sánchez-Dominguez JM, Montero E, Cañizares J, Bressani R, Escribá PV, Ruiz-Gutierrez V (2007) G protein-coupled receptor systems and their lipid environment in health disorders during aging. Biochim Biophys Acta 1768(4):964–975

Avila-Martin G, Galan-Arriero I, Gómez-Soriano J, Taylor J (2011) Treatment of rat spinal cord injury with the neurotrophic factor albumin-oleic acid: translational application for paralysis, spasticity and pain. PLoS ONE 6(10):e26107

Avila-Martin G, Galan-Arriero I, Ferrer-Donato A, Busquets X, Gomez-Soriano J, Escribá PV, Taylor J (2015) Oral 2-hydroxyoleic acid inhibits reflex hypersensitivity and open-field-induced anxiety after spared nerve injury. Eur J Pain 19(1):111–122

Avila-Martin G, Mata-Roig M, Galán-Arriero I, Taylor JS, Busquets X, Escribá PV (2017) Treatment with albumin-hydroxyoleic acid complex restores sensorimotor function in rats with spinal cord injury: efficacy and gene expression regulation. PLoS ONE 12(12):e0189151

Bento-Abreu A, Tabernero A, Medina JM (2007) Peroxisome proliferator-activated receptor-alpha is required for the neurotrophic effect of oleic acid in neurons. J Neurochem 103(3):871–881

Breuer S, Pech K, Buss A, Spitzer C, Ozols J, Hol EM, Heussen N, Noth J, Schwaiger FW, Schmitt AB (2004) Regulation of stearoyl-CoA desaturase-1 after central and peripheral nerve lesions. BMC Neurosci 5:15

DeWille JW, Farmer SJ (1992) Postnatal dietary fat influences mRNAS involved in myelination. Dev Neurosci 14(1):61–68

Galán-Arriero I, Serrano-Muñoz D, Gómez-Soriano J, Goicoechea C, Taylor J, Velasco A, Ávila-Martín G (2017) The role of Omega-3 and Omega-9 fatty acids for the treatment of neuropathic pain after neurotrauma. Biochimica et Biophysica Acta 1859(9 Pt B):1629–1635.

Hart CM, Tolson JK, Block ER (1991) Supplemental fatty acids alter lipid peroxidation and oxidant injury in endothelial cells. Am J Physiol 260(6 Pt 1):L481–L488

Hawkins RA, Sangster K, Arends MJ (1998) Apoptotic death of pancreatic cancer cells induced by polyunsaturated fatty acids varies with double bond number and involves an oxidative mechanism. J Pathol 185(1):61–70

Hostetler HA, Petrescu AD, Kier AB, Schroeder F (2005) Peroxisome proliferator-activated receptor alpha interacts with high affinity and is conformationally responsive to endogenous ligands. J Biol Chem 280(19):18667–18682

Kang JX, Leaf A (2000) Prevention of fatal cardiac arrhythmias by polyunsaturated fatty acids. Am J Clin Nutr 71(1 Suppl):202S-S207

Kien CL, Bunn JY, Tompkins CL, Dumas JA, Crain KI, Ebenstein DB, Koves TR, Muoio DM (2013) Substituting dietary monounsaturated fat for saturated fat is associated with increased daily physical activity and resting energy expenditure and with changes in mood. Am J Clin Nutr 97(4):689–697

Martínez J, Vögler O, Casas J, Barceló F, Alemany R, Prades J, Nagy T, Baamonde C, Kasprzyk PG, Terés S, Saus C, Escribá PV (2005) Membrane structure modulation, protein kinase C alpha activation, and anticancer activity of minerval. Mol Pharmacol 167(2):531–40.

Oh YT, Lee JY, Lee J, Lee J, Kim H, Yoon KS, Choe W, Kang I (2009) Oleic acid reduces lipopolysaccharide-induced expression of iNOS and COX-2 in BV2 murine microglial cells: possible involvement of reactive oxygen species, p38 MAPK, and IKK/NF-kappaB signaling pathways. Neurosci Lett 464(2):93–97

Pala V, Krogh V, Muti P, Chajès V, Riboli E, Micheli A, Saadatian M, Sieri S, Berrino F (2001) Erythrocyte membrane fatty acids and subsequent breast cancer: a prospective Italian study. J Natl Cancer Inst 93(14):1088–1095

Sales C, Oliviero F, Spinella P (2009) The Mediterranean diet model in inflammatory rheumatic diseases. Rheumatism 61(1):10–14

Sanchez-Bayle M, Gonzalez-Requejo A, Pelaez MJ, Morales MT, Asensio-Antonton J, Anton-Pacheco E (2008) A cross-sectional study of dietary habits and lipid profiles. The Rivas-Vaciamadrid study. Eur J Pediatr 167(2):149–154

Scholz J, Woolf CJ (2007) The neuropathic pain triad: neurons, immune cells and glia. Nat Neurosci 10(11):1361–1368

Scholz J, Abele A, Marian C, Häussler A, Herbert TA, Woolf CJ, Tegeder I (2008) Low-dose methotrexate reduces peripheral nerve injury-evoked spinal microglial activation and neuropathic pain behavior in rats. Pain 138(1):130–142

Scribe PV (2006) Membrane-lipid therapy: a new approach in molecular medicine. Trends Mol Med 12(1):34–43

Sekiya M, Yahagi N, Matsuzaka T, Najima Y, Nakakuki M, Nagai R, Ishibashi S, Osuga J, Yamada N, Shimano H (2003) Polyunsaturated fatty acids ameliorate hepatic steatosis in obese mice by SREBP-1 suppression. Hepatology 38(6):1529–1539

Stender S, Dyerberg J (2004) Influence of trans fatty acids on health. Ann Nutr Metab 48(2):61–66

Suresh Y, Das UN (2003) Long-chain polyunsaturated fatty acids and chemically induced diabetes mellitus. Effect of omega-3 fatty acids. Nutrition 19(3):213–28.

Troeger MB, Rafalowska U, Erecińska M (1984) Effect of oleate on neurotransmitter transport and other plasma membrane functions in rat brain synaptosomes. J Neurochem 42(6):1735–1742

Vögler O, López-Bellan A, Alemany R, Tofé S, González M, Quevedo J, Pereg V, Barceló F, Escriba PV (2008) Structure-effect relation of C18 long-chain fatty acids in the reduction of body weight in rats. Int J Obes (Lond) 32(3):464–473

**Gerardo Ávila Martín** holds a degree in Biology from Universidad Complutense de Madrid (UCM), a master's degree in the Study and Treatment of Pain, and a Ph.D. in Neurophysiology from Universidad Rey Juan Carlos (URJC). He has developed his career as a researcher in the Sensory-Motor Function Group of Hospital Nacional de Parapléjicos de Toledo (HNP). His study focused on the action of fatty acids in neuroprotection and neuroregeneration in animal models of spinal cord injury. He is currently working at the Health Integrated Area of Talavera de la Reina as a researcher in the Research Support Unit of Hospital General Universitario Nuestra Señora del Prado in Talavera de la Reina, Toledo. He teaches in the Interuniversity Master's Degree program in Clinical and Basic Aspects of Pain (Universidad de Cantabria-Universidad Rey Juan Carlos), and in different health research courses and training activities for residents in training and health personnel throughout the health area. He is a member of the Spanish Pain Society (SED). As a disseminator, he has experience in the participation and organization of dissemination activities such as *Science Week*. He is editor and reviewer of the journal of the Integrated Management Area of Talavera de la Reina (SALUX Journal).

Printed in the United States
by Baker & Taylor Publisher Services